Global Climate Change

A Clear Explanation and Pathway to Mitigation

David W Spitzer

Notice

The information presented in this publication is for the general education of the reader. Because neither the author nor the publisher has control over the use of the information by the reader, both the author and publisher disclaim any and all liability of any kind arising out of such use. The reader is expected to exercise sound professional judgment in using any of the information presented in a particular application.

Additionally, neither the author nor the publisher have investigated or considered the effect of any patents on the ability of the reader to use any of the information in a particular application. The reader is responsible for reviewing any possible patents that may affect any particular use of the information presented.

Any references to commercial products in this publication are cited as examples only. Neither the author nor the publisher endorses any referenced commercial product. Any trademarks or trade names referenced belong to the respective owners of the mark or name. Neither the author nor the publisher makes any representation regarding the availability of any referenced commercial product at any time. The manufacturer's specifications and instructions on the use of any commercial product must be followed at all times, even if they conflict with the information in this publication.

Because of the dynamic nature of the Internet, any web addresses or links contained in this book may have changed since publication and may no longer be valid.

ISBN: 9781932095203 (Paperback)
ISBN: 9781932095210 (Hard Cover)
ISBN: 9781932095227 (eBook)
Library of Congress Control Number: 2024911881

Previous Title: <u>Global Warming (aka Climate Change): An Understandable Data-Driven Explanation and Pathway to Mitigation</u>
Previous ISBN: 9798877871540 (Paperback)
Previous ISBN: 9798877878990 (Hard Cover)
Previous Library of Congress Control Number: 2024902261

This book was updated on 16July 2024.

Dedication

This book is dedicated to Ida, Deborah, Michael, Rebecca, Javier, and especially Harrison, Gabriel, Madeline, and Isabella, that they may ferret out facts and data, analyze them impartially, and think before reaching conclusions and acting.

Acknowledgements

The author acknowledges Kathleen Brush, Ph.D., MBA, Michael J. Furniss, and Ross Topliff for the time they spent providing sometimes brutal comments that were invaluable to writing this book. In addition, the author thanks Kathleen for providing encouragement, Michael for providing a broader perspective, and Ross for weeding out bias.

About the Author

David W Spitzer is an award-winning Professional Engineer and ISA Life Fellow who has written over 500 technical articles and 10 books. David's career has been dedicated to making industry more efficient and hence, more environmentally friendly. David currently consults, teaches, and performs forensic engineering.

About the Cover

Book covers in this genre often depict an industrial plant billowing large amounts of black or white smoke. The black smoke would be pollution, which was largely mitigated via regulation in the second half of the 1900s. The white smoke would be water vapor, which is environmentally benign. What cannot be seen is the colorless carbon dioxide in the emissions, which is a significant environmental concern.

The cover of this book reflects different outcomes that depend on whether appropriate actions to address carbon dioxide emissions are taken at appropriate times, or not.

Table of Contents

List of Figures and Tables

Preface

The genesis of this book was driven by my curiosity, and the frustration of seeing people with little or no technical background espouse information about climate change, the validity of which could not be readily determined. On the other hand, seemingly credible sources of information were not believable because their statements contained outright errors or, more often, errors of omission.

There are numerous books available about climate change and global warming that range from doomsday scenarios to the justification of the Green New Deal to denial. My quest for information did not uncover a comprehensive understandable source of information that utilized science and data to determine if global warming is real and, if so, provide a rational approach to mitigation.

To the contrary, virtually all the books were biased in some form, typically by misrepresenting data, making untrue statements, incorporating circular arguments, utilizing flawed information, committing errors of omission, and presenting opinion as fact. Scientific papers were typically focused on one aspect of climate change or global warming to the exclusion of other important factors. Curious and frustrated, I humbly set out to apply my technical knowledge and forensic experience to uncover the truth about climate change and wrote this book during the journey.

And what a journey it was. The original plan was to write about the science of burning fossil fuels and then analyze actual data to determine if global warming exists, what should be done about it, when it should be done, and gauge our progress to mitigate it. A pathway to mitigation was not in my original plan yet resulted logically from the analysis. Stated differently, there were no preconceived notions of problems or solutions prior to writing this book.

Research revealed that discovering the role and importance of global warming in context involves understanding life before fossil fuels were burned in quantity and recognizing that our response to global warming has been largely shaped by seemingly incompatible worldviews. The first few chapters, which describe several general concepts and life around 1800, and the final chapters, which suggest that global warming can be part of a larger agenda, became important afterthoughts of researching and writing the middle of the book, which shows that the accumulation of carbon dioxide in the atmosphere is a more immediate concern. Despite best efforts and much hoopla, no previous actions have addressed this accumulation aggressively. However, it appears that technology will soon be available to start meaningful mitigation and, at the same time, slow, stop, and reverse global warming. Read on to find out why and how.

<u>Introduction</u>

*A fast way to get to the bottom of a mountain is to jump off a cliff. ---
David W Spitzer*

Climate change is no laughing matter because it can profoundly affect you, me, and every other living organism on Earth. Yet there is considerable debate about the issue and, much more importantly, what should be done about it. The stakes can be catastrophic when incorrect actions are taken or when correct actions are taken at the wrong time.

Some people claim that burning fossil fuels causes the accumulation of carbon dioxide in the atmosphere, which warms the Earth. This represents an existential climate crisis, so humans should stop burning fossil fuels and immediately pivot to renewable energy sources.

Others claim that there is no immediate existential climate crisis because carbon dioxide is a trace gas that has not been proven to appreciably affect the temperature of the Earth.

What we have here is no less than an existential crisis that is not an existential crisis, and global warming that is occurring and is not occurring. Amazingly, these two diametrically opposite conclusions were presumably reached by educated people who had access to essentially the same information.

As is often the case, the truth lies somewhere in between, because each group presents its own truth, but not the whole truth. Nonetheless, the very existence of these two diverse opinions screams for an objective investigation driven by data, and not by ideology or preconceived notions.

This book analyzes the intersection of human evolution, scientific advancements, demographics, population growth, combustion, energy consumption, greenhouse gases, and meteorological data to determine if global warming exists and, if it does, to put it in perspective before identifying trends, proposing actions, and quantifying results. By providing a working knowledge of global warming and its manifestation in daily life, this book will help the reader better understand what is said and the results of actions taken.

Instead of presenting complex calculations, data is presented largely in graphical form that should be understandable to laypeople and informative for those with a technical background. In addition, this book provides an understanding of many related subjects and tools that can help the reader analyze information and claims about global warming and climate change.

Disclaimer: Knowledge is constantly evolving, so the author reserves the right to supplement the data and analysis herein should additional information become available.

Chapter 1: Tools of the Trade

A bad workman blames his tools. --- Proverb

Did you know that I caused the northeast blackout in November 1965 that plunged over 30 million people into darkness for about 13 hours in the northeastern United States and part of Canada? I was at home, minding my own business, just as it was getting dark outside. I turned on the light above my bed, and suddenly, all the lights in the apartment went out.

I went outside only to discover that the lights were out as far as I could see. The traffic lights were not working, so random people manned intersections to direct traffic. Back in my room, my little transistor radio was able to receive local broadcasts that slowly began reporting the extent of the outage. What they did not report was that it was I who caused the entire outage by turning on the light above my bed. After all, the lights went out just as I turned on my light, so my light must have been the reason that the blackout occurred.

As you might have suspected, the actual cause of the blackout was not my light turning on but rather a chain reaction that started when a mis-programmed safety relay tripped.

---xxx---

It is well known that no person can know everything about everything. However, people can do research and get information about almost everything. Solving a problem involves defining what information is needed, determining how to obtain that information, and pragmatically assessing how much detail is required about that information. Implementing these actions often involves more art than science.

For example, the solution to a problem might entail knowing the carbon dioxide concentration in the atmosphere 100,000 years ago. An investigation will reveal that this information is contained in air bubbles that were frozen in ice 100,000 years ago in Antarctica. An investigator could learn all about ice core technology and dig ice core samples in Antarctica. However, it would be faster, easier, and less expensive to locate the results of ice core sample experiments and read the concentration from a graph. In other words, it is assumed that the people who performed the ice core experiments and developed the graph were experts in the field, knew what they were doing, and generated the graph correctly. Therefore, a reasonable investigator can accept the concentration information and does not have to thoroughly research ice sample technology and practices, with the important caveat that questionable data or new information may require a deeper investigation and a reevaluation of the results based on the findings.

For example, global warming may have adversely affected the Arctic polar bear population. Fundamental to this analysis is data relating to the past and present Arctic polar bear populations to determine whether the Arctic polar bear population is increasing, decreasing, stable, or something else. A polar bear organization published information that indicates that the population is decreasing. However, further investigation revealed that it is virtually impossible to determine the past or present populations. In addition, there is a book that shows that the population is increasing. The takeaway is that no one really knows, so the information is suspect and should not be used for analysis.

In general, further research is warranted when information seems suspect or when the subject just happens to be of personal interest.

Addressing Global Warming

Of immediate interest are the answers to the following questions.

- Does global warming exist?
- To what extent has global warming progressed?
- What should be done (if anything) about global warming?

Selecting where to start is paramount because the *perceived* nature of events can change when analyzed from different vantage points. The Earth has been warming, cooling, and evolving since its formation approximately 4.5 billion years ago. [1] The Earth's birthday seems like a logical place to start. However, there was no life on Earth at the time.

Approximately 200 million years ago, the concentration of carbon dioxide in the atmosphere was four times its current concentration. Another spike occurred approximately 50 million years ago. This data gives the impression that the current concentration should not be an issue because the Earth has already survived high concentrations. However, these data points might not be all that relevant, because the Earth had different flora, fauna, and geography millions of years ago.

Burning fossil fuels releases carbon dioxide into the atmosphere. The concentration of carbon dioxide in the atmosphere started to increase considerably around 1950. The temperature of the Earth started increasing steadily in 1977. Whatever caused these changes had been occurring for some time, so 1950 is too late to start an analysis of global warming.

Steam engines were invented in the early 1700s but were not in wide use until around 1800, when the burning of fossil fuels to operate machines became more common, thereby increasing fossil fuel consumption. This looks like a good time to start an analysis.

Climate change and global warming cannot be fully understood and adequately addressed in a vacuum, because it is effectively the price that humans are paying for enabling the global population to expand and live a 'life of life' instead of the 'life of death' that was the human condition for thousands of years and is still the human condition for approximately 10 percent of the Earth's population, as will be more fully discussed in Chapter 4.

Spoiler Alert: Chapters 3, 4, and 5, chronicling various aspects of this amazing human journey that started around 1800, are not about climate change per se but have everything to do with human progress and how the human condition will be adversely affected if incorrect actions are taken or if correct actions are implemented too quickly. Chapter 2 presents some tools that can be used to identify suspect information and help obtain additional information so that analyses can be based on the best available data, and not flawed information, errors of omission, or opinions.

The combustion of fossil fuels, such as coal, gas, and oil, generates carbon dioxide that is accumulating in the atmosphere. A working knowledge of combustion and energy consumption processes is fundamental to understanding this process and its mitigation. Chapters 6, 7, and 8 explain the fundamentals of energy sources, combustion, electricity, and end-uses to enable the reader to understand how global warming is rooted in science, and not in emotion.

Chapters 9, 10, and 11 discuss greenhouse gases, how they affect the Earth, international agreements, and disagreements therewith. Chapter 12 presents broad strategies to reduce carbon dioxide emissions that are more fully developed in Chapters 13, 14, 15, and 16 for electricity generation, transportation, industrial, and residential and commercial energy end-use sectors, respectively. Chapter 17 presents suggestions to reduce the carbon dioxide concentration in the atmosphere that are quantified in Chapter 18. Chapter 19 provides perspective on where the Earth is in its journey to a lower carbon future, while Chapter 20 addresses some of the politics of climate change. Importantly, Chapters 21 and 22 investigate how climate change and global warming fit into a larger agenda.

The underlying data for the above analysis is presented graphically, sometimes with different granularity to show causality, coincidence, or no correlation between parameters.

Coincidence and Causality

In the previous example, an increasing Arctic polar bear population would not support the premise that Arctic polar bears are being adversely affected by global warming. However, a decreasing population might just be a coincidence caused by other influences.

All too often, people conflate events that occur coincidentally and conclude that one caused the other. For example, global population growth started around 1800. What happened in 1800? For starters, Napoleon marched into Austria, the Library of Congress was founded, the United States census was released, John Adams became the first president to live in the White House, and the United States Congress held its first session in Washington, DC.

Given these choices, one could opine that the existence of the Library of Congress, the issuance of the United States census, and where John Adams lives should not affect the global population. Napoleon's march into Austria would tend to reduce the population, but Austria had relatively few people. Therefore, it must be the location of the United States Congress meeting that caused the global population to increase, just like me causing the northeast blackout. This conclusion does not make sense, but a tortuous and logically challenged case could be made that Washington is at fault, and it all started when Congress moved. Some legitimate explanations can be complex, but beware of explanations that are overly complicated, tortuous, or use questionable logic.

One case in point is the original statement that something happened *around* 1800. However, the events listed occurred *in* 1800. This small, seemingly insignificant word change (that most people would not notice) eliminates many important events and processes that occurred during the years, decades, and centuries *around* 1800 that are related to the increasing global population. One must be careful to ensure that issues and analyses are accurately described. It helps to listen attentively, read slowly, and write carefully.

It is extremely important to understand the difference between coincidence and causality. Just because two events happen at the same time does not mean that one caused the other. It may just be a matter of coincidence. However, when one event follows similar events consistently, the two events may be linked by causality. For example, it is not a coincidence that looting tends to increase during blackouts and subside shortly after electricity is restored.

Implying causality from coincidental events is rampant. People and the media often conflate events to promote their own viewpoints, as appears to be the case with the Arctic polar bear. Similarly, information and data related to climate change and global warming are rarely presented to the public fairly, completely, coherently, and in context.

Granularity

The measure of the amount of detail present in data is its granularity, where data containing more detail is more granular. The level of granularity that is useful is different in different circumstances. Data may be required every day, month, year, decade, century, or other period, depending on the issue being analyzed. For example, a granularity of millions of years might be sufficient to analyze information about dinosaurs, while annual data is appropriate for analyzing recent global population trends. Faster processes can require data every few minutes to perform a valid analysis.

Selecting the appropriate granularity is often more a matter of art than science. Too much granularity can make analysis cumbersome and difficult, whereas analysis may not be possible when the available data has too little granularity.

Incongruent Occurrences

It is not uncommon to analyze a situation and determine that two seemingly opposite forces occur at the same time, such as something bad that is also improving at the same time. Bad refers to the current state, while improving indicates that the current state is changing for the better. For example, a person might be admitted to an intensive care unit due to a lung injury caused by an automobile accident but two days later the person is breathing better and requires much less oxygen. The situation is still bad, but the person's health is improving.

Approximately 10 percent of the world's population currently lives in extreme poverty. It is terrible to describe the lives of almost 800 million people in this way. However, as we will discover, almost 80 percent of the world's population in 1800 lived in extreme poverty. Things may be bad now, but the lives of billions of people have improved tremendously in just two centuries.

Similarly, things can be true and irrelevant at the same time. In the previous example, start with the premise that one person in poverty is too many. We can agree on this because we do not want anyone to live in poverty. However, starting with this premise is unrealistic because we all know that there will always be someone in poverty somewhere. Therefore, the mantra 'one person in poverty is too many' is irrelevant and tends to stifle meaningful discussion. If allowed to continue, such narratives can become distorted in a manner that may be self-serving and result in actions that are a detriment to the greater good.

Forensic Analysis

Most analysis in our daily lives is performed by starting with a certain event and then determining how the consequences of that event affect the future. For example, turning on an oven will cause the oven to get hot, cook food, and produce carbon dioxide that will be dispersed into the atmosphere. Note that the events are in chronological order in the forward direction.

Much of the analysis in this book is forensic and fundamentally based on artifacts that are available now but that can be used to determine what occurred in the past. In other words, the analysis is performed backwards to determine what occurred in the past based on the limited amount of information that is available now. For example, the amount of carbon dioxide in the atmosphere is currently measured with analyzers that have only recently been developed. However, scientists use techniques such as analyzing ice samples to determine how much carbon dioxide was in the atmosphere hundreds of thousands of years ago.

Forensic engineering can also be applied to processes. For example, population data combined with other seemingly unrelated data will help the reader understand why the global population started growing around 1900, with the intent of understanding and determining approximately how long this growth will continue. Further, carbon dioxide emissions will increase to support the growing population, so knowing when the global population will stabilize is one factor that can help influence how quickly mitigation should proceed.

Graphs and Tables

Graphs are used throughout this book to present data and show trends. Their meaning and significance are explained, so the reader will develop a deeper understanding of the data being presented and become comfortable examining the graphs.

Further, it is extremely important to consider all the information present on the graph in the context of the process that is occurring. This book will present graphs with explanations as to how examining only part of the information can lead to incorrect conclusions.

Graphs of processes can take many different shapes. Some processes follow a straight line. However, some processes encountered in nature start at one value, rise rapidly, and then stabilize at a higher value. For example, the height of a boy is approximately 50, 90, and 180 centimeters at birth, when he is 2, and 20 years old, respectively. Examining growth only during the first two years of life would lead one to believe that boys grow 40 centimeters every two years. If so, the boy would be 130, 170, 210, and 250 centimeters tall at ages 4, 6, 8, and 10 years old, respectively. Adult men do not grow to be 250 centimeters tall, and neither does a 10-year-old boy. The actual growth process is not linear but rather exhibits an S-curve with rapid growth at first before slowing down and stabilizing at the boy's adult height. In this process, the boy grew 40 centimeters in just two years, but then growth slowed, and it took him another 18 years to grow another 90 centimeters.

Data is sometimes presented in the form of a table. The meaning and significance of the information contained in tables are explained in the text. Examining the actual table entries will impart a deeper understanding of the data.

Previous Environmental Successes

The quality of the air we breathe may or may not have been pristine before 1800, but increased factory operations significantly degraded air quality by the 1940s.

In October 1948, Donora, Pa., was enveloped in a lethal haze. Over five days, nearly half of the town's 14,000 residents experienced severe respiratory or cardiovascular problems. It was difficult to breathe. The death toll rose to nearly 40. Disturbing photos show Donora's streets hidden under a thick blanket of gray smog. A warm air pocket had passed high above the town, trapping cooler air below and sealing in pollutants. Donora was no stranger to pollution. Steel and zinc smelters had long plagued the town with dirty air. But the air pocket left pollutants with no escape route. They sat stewing in the streets, where residents breathed them in lethal doses.

The situation in Donora was extreme, but it reflected a trend. Air pollution had become a harsh consequence of industrial growth across the country and world. Crises like Donora's were widely publicized; people took notice and began to act. Scientists started investigating the link between air pollution and health. States began passing legislation to reduce air pollution. And in 1970, a milestone year, Congress passed the Clean Air Act Amendments which led to the establishment of the nation's air quality standards. [2]

What and how much should be regulated can be debated, but the fact remains that humans made deliberate changes that improved global air quality. The Federal Water Pollution Control Act of 1948 and subsequent amendments that set standards for water quality similarly addressed water pollution in the United States and helped build many water and wastewater treatment plants where none had previously existed. This begs the question of where the sewage was going prior to building the wastewater treatment plants. One answer is: into a river via an open-air sewage ditch. [3] Stated differently, the environment was not necessarily as pristine as we would like to believe.

Research in the 1980s discovered that the ozone layer surrounding the Earth was extremely depleted.

> Ozone is a colorless gas mainly found in Earth's stratosphere. It forms a protective layer that absorbs harmful ultraviolet light from the sun. In 1985, an extreme depletion of ozone over Antarctica was discovered—the so-called Antarctic ozone hole. It soon became clear that this drop in ozone was caused by man-made chemicals called chlorofluorocarbons (CFCs). To help solve the global depletion of ozone, the international community regulated CFC production and consumption by adopting the Montreal Protocol in 1987. [4]

In 2023, a United Nations panel of experts reported that the Earth's ozone layer was on track to recover by 2060. [5] In this case, the world came together and acted globally to address the problem.

A significant example of humans coming together to reduce electrical energy consumption, which in turn reduces the production of carbon dioxide, is the adoption of LED light bulb technology. Incandescent light bulbs, developed by Thomas Edison in the late 1800s, were phased out starting in the 2010s after more than a century of use. Legislation was passed to replace them with LED light bulbs that last much longer, produce the same amount of light, and use approximately one-sixth of the electricity. It was predicted that, simply shifting to more efficient lighting technologies in all lighting applications would:

11

- Save electricity equivalent to the output of more than 250 large coal-fired power plants.
- Reduce global electricity consumption by approximately 5%.
- Reduce annual CO_2 emissions by at least 490 million tons.
- Be equivalent to taking more than 140 million mid-size cars off the road.
- Avoid the construction of 252 power plants, which is equivalent to US $210 billion in investments saved. [6]

In another example, communicating wirelessly with the Internet provided faster, more versatile, and more convenient communications than hardwired connections while requiring less metal (resources) than their hard-wired predecessors. On a side note, it also eliminated conflict among family members about how to share the much slower dial-up Internet connection.

Note that the implementations used to address these problems were quite different. The air and water quality became so untenable that something had to be done. Companies and communities were unwilling to act, so legislation was passed that forced the installation of pollution abatement equipment to resolve the problem, which resulted in clean air and water. Research indicated that the ozone layer was decreasing. A global agreement was reached to eliminate CFCs, and the ozone layer was put on a path to recovery. Significantly more efficient lighting was developed that was not wholeheartedly embraced by the manufacturers who profited from selling replacement incandescent light bulbs. Legislation was passed to force elimination of incandescent light bulbs over a reasonable time frame, which reduced the demand for electricity and carbon dioxide emissions. The transition from hard-wired to wireless Internet connections occurred quickly because it was demonstratively so much better.

The nature of these solutions is particularly interesting. Air and water quality legislation was implemented because the situation was dire. While not legislation per se, virtually all countries agreed to eliminate CFCs because a dire situation was credibly predicted. Legislation was passed to accelerate the adoption of LED light bulbs. The transition from wired to wireless Internet communication occurred organically.

In general, organic implementations that improve the well-being of the population are widely and willingly accepted by the population. However, few global problems have organic solutions, and if they did, the problem would have been resolved long ago. Forced implementation, such as using legislation or agreement, often garners opposition from the population, or from groups within the population.

Some problems, such as addressing air and water quality, required stringent prescriptive legislation. Increasing the ozone layer was implemented by pseudo-legislation and directly affected only a small segment of industry. Adoption of industry standards has the potential to be another form of pseudo-legislation. Legislation to replace light bulbs was enacted when LED light bulb technology was sufficiently mature.

All these implementations were successful, and any one of them could be chosen as the preferred model to address future problems. However, organic solutions to large problems are rare. Letting a situation become dire before acting is not prudent. Therefore, legislation and pseudo-legislation are sometimes necessary, but can cause unintended damage and acrimony in the population. Legislation or pseudo-legislation transparent to the population, involves a small sector of industry, is feasible, and allows reasonable time for implementation would tend to have better chances for success.

Of these examples, the implementation of replacing incandescent light bulbs with LED light bulbs provides a pragmatic model to address future problems. The implementation was transparent to the population because LED light bulbs had similar appearance, form, and function, where only the manufacturers needed to change their products. **Importantly, the technology was sufficiently mature, and the legislation allowed sufficient time for the manufacturers to adapt.** Legislating a solution utilizing unproven and insufficiently developed technology is a formula for failure. For example, California's mandate to require automakers to provide electric vehicles was repealed in 1996. [7] It would be fair to surmise that the mass production of electric vehicles was not viable at that time. It took another 25 years or so for the technology to mature and capture a small, but meaningful share of the market.

Countries have demonstrated the ability to act in harmony to address other global issues, such as protecting endangered species from extinction and land masses from destruction, so it is certainly

13

possible that this can be repeated to address climate change. Science should be used to determine the actions that should be taken. However, common sense and pragmatic thinking should determine the speed at which actions should be implemented to avoid negative results and unintended consequences.

Chapter 2: The Search for Facts and Data

Believe nothing you hear, and only one half that you see. --- Edgar Allan Poe

All kinds of people with all kinds of experiences exist in all kinds of places. Some people use facts, whereas others are more emotional.

Some years ago, after an introduction and brief conversation, a gentleman who happened to work for the United Nations replied quickly to my inquiry about his work with the premise that the United Nations must help people. He continued explaining his work to help people in a particular country before concluding that the United Nations must help people.

Analyzing the country's issues on their merits may very well conclude that these people should get help, which could include various levels of active support or simply standing by ready to help only if needed.

Interestingly, the man embedded his conclusion into his original premise, thus forming the basis to justify his conclusion. Impeccable! By starting with his conclusion as a fact, this circular argument will always be correct, and he cannot lose.

---xxx---

By observation, different people use different processes to understand and analyze issues that might have multiple solutions, or sometimes no solutions at all. Further, some of the processes employed to define and analyze problems may be flawed, leading to counterproductive solutions that can worsen the problem, often catastrophically. For example, carbon dioxide emissions emitted by comparable hybrid and electric vehicles might be similar. However, electric vehicles are touted as clean, while hybrid vehicles are not. In addition, the higher mineral content of electric vehicles promotes more human and environmental abuse in the countries in which they are mined. Therefore, a hybrid vehicle would emit less carbon dioxide.

It is particularly important to address global issues objectively based on facts, because potential solutions can adversely affect almost every aspect of human existence. Climate change fits this bill perfectly, which is why it is so important that it not be addressed emotionally.

Facts and Data

Facts are statements that are not disputable. They are based on empirical research, quantifiable, and agreed to by a consensus of experts. Facts are more than theory and can be proven through experience and calculation. Some facts can be obvious from the observation of an event that occurred in the past. Facts are not opinions or theories. An accepted truth is not a fact.

Facts are fundamental to analyzing issues and making effective decisions. Failure to understand and consider all the facts can be catastrophic. For example, electric vehicles produce no emissions. This is a fact that can be used to imply that all vehicles should be electric. Full speed ahead. **No, not so fast!** Electric vehicles contain materials whose manufacturing processes produce emissions, are manufactured in processes that produce emissions, and operate using electricity whose generation produces emissions. All emissions should be considered to determine if the total carbon dioxide emissions associated with an electric vehicle are lower than those of a comparable gasoline, diesel, or hybrid vehicle. Failure to consider all emissions can result in the purchase of an electric vehicle that is expensive, inconvenient to operate over long distances, and may not produce the lowest emissions as compared to (say) hybrid vehicles that can be fueled at ubiquitous gas stations without installing a new infrastructure for charging.

Data is factual information, such as measurements, that is used as a basis for reasoning, discussion, or calculation. [8] For example, consider an analyzer that measures 420 parts per million (ppm) of carbon dioxide in ambient outside air. After observing that the analyzer was properly installed, calibrated, and operated, multiple experts would agree that there was 420 ppm carbon dioxide in the ambient air at that location at that time. This data can be used to analyze concentration trends.

Some data is more reliable than others. For example, carbon dioxide in the air is measured and reported daily, so historical measurements are available. However, putting recent measurements into perspective might benefit from knowing how much carbon dioxide was in the air one hundred thousand, one million, and one billion years ago. These values cannot be obtained from actual measurements because analyzers did not exist.

Experts have developed methods to infer past values. However, there is generally less confidence in values farther back in time. In addition, there is generally more confidence in the information when multiple scientific methods can be used to confirm the value of a parameter at the same time.

Inferred past measurements are used throughout this book because they are the best information available. For example, graphs showing the carbon dioxide concentration in the atmosphere 400 million, 10 thousand, and 2 thousand years ago will be analyzed. They were not measured directly but rather inferred using other scientific techniques.

Efforts have been made to ensure that the data and graphs presented herein represent actual measurements or the work of experts competent in their respective fields that have been peer reviewed and generally accepted. Therefore, it is usually not required to know or understand the details of the science and methodology behind the data. However, a more detailed scientific investigation into the technology and methodology is warranted when the data appears to be suspect.

Although not related to climate change, a cursory examination of a legal case illustrates how asking inconsequential questions can lead virtually nowhere, whereas asking focused questions, performing research, and examining facts and data can identify a meaningful path forward. **The importance of this example should not be underestimated, as illustrated in the following example that suggests an approach that can potentially transform a process rife with contentious opinions into a process based on an analysis of facts and data.**

In this case, the defendant was prosecuted for allegedly overvaluing multiple assets to secure favorable loan terms, for which the suggested punishment was to bar the defendant from doing business within the jurisdiction plus damages of USD 250 million, [9] which was subsequently increased to over USD 370 million. [10]

In a ruling made prior to hearing a defense, the defendant was declared guilty of grossly overvalued multiple properties, including his residence, estimating it was worth as much as $612.1 million, though an assessor said (according to the judge) that its market value was no more than $27.6 million, [11] and his previous residence, where its size was nearly three times its actual size. [12] The cancellation of certificates was ordered along with the appointment of a receiver to manage the dissolution of the defendant's businesses. [13] The defendant countered that assessed values were incorrectly used to value assets, instead of market values, and the incorrect size was an error, respectively. What is your position on this outcome based upon this limited amount of information?

As a factual matter, appropriate legal procedures were followed, and the ruling was against the defendant. On the other hand, is it your opinion that defense arguments should have been heard before ruling on such a consequential matter? Why or why not? Do you believe that the prosecutor accurately valued these assets? Why or why not? Do you believe the explanations made by the defendant? Why or why not? The judge later wrote that all the defense expert testimony proves is that for a million or so dollars, some experts will say whatever you want them to say. [14] Does this change your opinion? Why or why not? [A]

The answers to these questions might elicit opinions and perhaps reveal biases, but they will likely not yield a reasoned position on the matter. Perhaps it would be more appropriate to remove opinion and bias by comparing the different valuations to the actual selling price of property in the same area at the time in question. There appear to be no properties of comparable size to the defendant's residence, however an internet search revealed that a nearby oceanside mansion sold for approximately USD 12.5 million in 2009 and 2019. [15] The property cited by the judge is a National Historical Landmark that is approximately 10 times larger and sits on approximately 20 times more land with both an oceanfront (not oceanside) and lakefront. [16] The original owner spent USD 7 million on the property in the 1920s, which was equivalent to over USD 90 million in 2017. [17] A magazine report published months before the indictment suggests that the property was worth approximately USD

[A] This judge's statement regarding the expert might seem outrageous, but as I approached the witness stand to be sworn in to testify in my first legal case, the judge said that *you can buy an expert to say anything*. Seated in the elevated witness stand beside the judge's bench in a subsequent trial, I was able to observe a different judge reading a newspaper during my testimony.

170 million in 2020. [18] Has your position on the outcome of this case changed? Why or why not?

Specifically included because it is not related to climate change, this example illustrates how asking focused questions, performing research, and examining facts and data can help identify a meaningful path forward while ignoring other factors that are not relevant. **Many aspects of climate change presented in this book follow this pattern.**

Factfulness

Hans Rosling (1948-2017) was a Swedish physician and professor of international health who, among many other accomplishments, identified ten instincts that are often used to manipulate the meaning of information. Using one or more instincts may be an innocent oversight or intentionally introduced to be misleading. Either way, identifying when the instincts are used can help better understand the information presented, identify information that is lacking, formulate probing questions to ferret out facts, and can be extremely useful in virtually all aspects of life, such as health, science, relationships, family, and business, to name a few. The following is a summary of the instincts. [19]

- The *gap instinct* is the first and most powerful instinct. It implies that one group is different than another, when there may be little difference between the groups. For example, one group of students averaging ten points higher than another on an exam implies that one group is superior. However, graphing the distribution by score may well indicate that the number of students obtaining the same score is virtually the same in both groups.

   Another part of the *gap instinct* is to present only one number or only one data point. This should **immediately raise a flag to investigate further**, because that one number provides no point of reference for comparison and analysis. For example, the carbon dioxide level measured at the Mauna Loa Observatory (Hawaii, USA) was approximately 420 ppm. Wow, imagine that! Let's get excited and mobilize! But seriously, how can you determine if 420 ppm is high, low, good, bad, or normal without a point of reference or trend data with which to compare? A reasoned determination cannot be made without additional information.

- The *negativity instinct* warns us to expect to hear bad news, because people pay attention to sensational events that are usually bad and ignore good news that is usually boring. For example, warnings of impending disasters due to climate change and global warming abound, whereas the United States reduction of its per capita carbon dioxide emissions by approximately one-third goes virtually unnoticed.
- The *straight-line instinct* warns us not to assume that a trend will continue as a straight line. For example, the concentration of carbon dioxide in the atmosphere increases in the Northern Hemisphere summer and decreases in the winter. Using average monthly measurements to extrapolate future concentrations as a straight line will yield erroneous results.
- The *fear instinct* warns us to beware of overreacting when there is the possibility that something bad will happen by introducing fear. **The fear might be real, but the chance of a bad outcome may be small.** For example, it was reported that "people are now suffering and dying … because heat waves, droughts, floods, wildfires, disease outbreaks and other dire effects of climate change are accelerating." [20] Is this completely true? If so, where is it occurring, why is it occurring, and **what are the chances that it will happen to me**?
- The *size instinct* alerts us to put things in proportion. For example, the previous example not only introduces fear, but also elevates the size of that fear out of proportion for an overwhelming majority of the population.
- The *generalization instinct* warns us not to generalize about groups of people or objects. For example, electric vehicles are categorized as zero-emission vehicles. However, their manufacture, operation, and recycling activities emit significant amounts of carbon dioxide that may exceed those of a comparable hybrid vehicle that emits carbon dioxide.
- The *destiny instinct* warns us not to pigeonhole people, groups, or things. For example, stating that a certain group will not change does not recognize that slow change still

represents change and that the group could catch up or achieve a given goal.

- The *single instinct* teaches us not to expect to use the same tool for every task, but rather to have multiple tools at your disposal from which you can choose the best tool for the job at hand. For example, solar energy production is heavily promoted and might make sense in the vast desert near Las Vegas (Nevada), but not necessarily in central London, (United Kingdom), which receives less energy due to its higher latitude and is known for its fog, which blocks much of the available solar energy.

- The *blame instinct* instructs us to resist pointing the finger at others when something goes wrong. Learn from what happened and move forward. For example, my predecessor at the plant tried to combust waste fumes in the main flame of an incinerator to reduce fuel consumption, increase capacity, reduce chemical emissions, and reduce carbon dioxide emissions. My success where he failed was a direct result of my learning from what went wrong. Frankly, my success would not have come if he had not tried and failed.

- The *urgency instinct* **warns us not to act rashly,** but rather to take small, measured steps to achieve a goal. For example, immediately banning the use of all fossil fuels will reduce the concentration of carbon dioxide in the atmosphere, reduce global warming, and create a multitude of insurmountable problems. Reducing the use of fossil fuels should be addressed in small increments.

Factfulness in Action

Factfulness instincts abound in almost every facet of everyday life. For example, consider these two sentences from the Introduction, repeated here for convenience.

Some people claim that burning fossil fuels causes the accumulation of carbon dioxide in the atmosphere, which warms the Earth. This represents an existential climate crisis, so humans should stop burning fossil fuels and immediately pivot to renewable energy sources.

Table 2-1 provides comments regarding the applicability of each instinct to this text. These comments are intended to reflect the content of the text without regard to whether their substance is correct or applicable.

Table 2-1 Comments on Rosling Instincts Applied to Text

Instincts	Comments
Gap	There is a difference (gap) between historical and current levels of carbon dioxide in the atmosphere.
Negativity	The accumulation of carbon dioxide in the atmosphere is bad. Global warming is bad.
Straight-Line	Global warming will continue unabated if nothing is done.
Fear	Global warming is an existential climate crisis.
Size	Global warming is a global issue.
Generalization	No data is presented.
Destiny	Global warming will continue unabated if nothing is done.
Single	Stopping the burning of fossil fuels and pivoting to renewable energy sources is the only solution.
Blame	Society is at fault.
Urgency	We must act immediately.

This exercise revealed that the two sentences in the above text include all ten instincts! It is no wonder that this message has received such widespread acceptance.

Circular Arguments

Circular arguments return to their origins without proving anything. They typically start by stating conclusions as *facts*, using the *facts* in discussion, and then concluding that the starting conclusion is valid. Circular arguments are ubiquitous, as exemplified by the story at the start of this chapter. In general, it is difficult to lose an argument when you start with the conclusion.

Flawed Information

Information can be manipulated to produce a desired result, which can be nefarious. If there are any doubts about this, pick a somewhat controversial event, listen closely, and evaluate how it is communicated by different people and different media outlets. The event should be reported completely and objectively. However, the reports are often different. Some media outlets and people may decline to report about the event, not acknowledge that the event occurred, actively censor the existence of the event, or flat out lie about the event. Many flaws are the result of unintentional oversights and mistakes, whereas other flaws are intentional. Recognizing and correcting flaws can be difficult, especially when verifiable information is not readily available.

For example, reporters are consciously or subconsciously incentivized not to report good news because it is typically boring. In contrast, bad news is usually more interesting and tends to increase readership, viewership, and advertising revenue. To wit, reporting about a picture of a poor, lonely, hungry polar bear fishing from an Arctic iceberg, allegedly the result of global warming, would be more exciting and newsworthy than reporting that every commercial aircraft took off and landed safely last year, while carrying billions of passengers. This good news is more impactful but is rarely reported, while the bad news is often highlighted.

Governments, organizations, and individuals with an agenda often convey a distorted perspective of reality by committing errors of omission, or by stating opinions that appear to be facts. Reporters often spin stories to support their agenda, so that certain events and people are elevated or demonized to support a favored ideology. Scientific information can also be similarly spun, such as when using a graph to forecast despite a disclaimer stating that the graph is not a forecast.

For example, as a factual matter, the above-described scene involving the poor, lonely, hungry polar bear fishing from an Arctic iceberg is perfectly normal and could have occurred 5 days ago or 500 years ago. Regardless of the facts, appropriate music, words, and text can be selected to support an agenda, or not.

Everyone does this subconsciously to some extent. Understanding these and other instincts of communication will bring individuals and institutions closer to the truth, and result in better decisions.

Errors of Omission

So-called facts can have insidious flaws. By way of example, a few years ago, media outlets reported that an official claimed that his actions in a particular matter were performed in accordance with federal regulations. Two media outlets with opposing political views published almost identical articles on the same day, suggesting that the articles had a common source. Both articles factually reported what the official said and contained errors of omission. One article contained a link to a federal regulation that supported that publication's narrative, revealing that the actions were not in conformance. The link did not appear in the other article, and neither article questioned the validity of the claim. In summary, both articles described the event accurately. However, both contained errors of omission that led readers to believe that the politician's claim was valid when it was not.

This pattern illustrates both laziness and an injection of an agenda into the news. Laziness, in the sense that the author of the article with the link did not read the regulation or did not understand it sufficiently to verify the validity of the claim, which, as it turned out, supported that author's agenda. Injection of a political agenda, in the sense that the absence of the link denied readers the opportunity to investigate and discover that the claim was false. Stated differently, details were not the official's friend, but rather the reader's enemy.

Accurate reporting of an event can be factual. However, further investigation is often necessary to understand the significance of that event in context. Multiple facts can occur at the same time, such as when an event occurs, someone makes an incorrect statement about the event, and the incorrect statement is reported. The report factually describes the statement made. However, the validity of the statement is not addressed. The lesson to be learned is that no one can be trusted, so you need to **investigate for yourself**.

Progressive Disclosure

Most people can readily explain the ins and outs of their area of expertise, but being pragmatic, they often limit their answers and explanations to the basics, lest they talk for hours while the listener politely smiles to express interest. Basic answers are usually reasonable because the listener will request more detail if required. The downside is that the listener often does not know that more details are needed.

For example, the accuracy of a carbon dioxide analyzer would be compared with the accuracy of similar analyzers to decide which analyzer to purchase. However, the analyzer will be operated outdoors in the hot sun, rain, and snow, so how do ambient conditions affect its accuracy? The incoming power varies from 90 to 130 volts, so how does this affect its accuracy? How and how often does the analyzer need to be calibrated?

The overall concept is that each question begets another question, progressively disclosing additional information until the questioner is satisfied. This can create an insidious issue because information can be intentionally withheld when the questioner does not know what questions to ask or does not have access to the answers. Worse yet, environmental issues involve many wide-ranging topics, so it can be difficult to know what questions should be asked. This can cause poor decisions to be made.

Consider the previous example, where the carbon dioxide level measured at the Mauna Loa Observatory (Hawaii, USA) was approximately 420 ppm. Lay people may sense a gut reaction to the number, but most will not know to ask if 420 ppm is high, low, good, bad, normal, or something else. Many would see 420 ppm as something that needs to be fixed now, instead of waiting for a few years until better technology becomes available.

Opinions

Opinions are statements that represent a judgment or viewpoint, typically of a person or organization. Opinions may or may not be based on fact, may or may not be technically valid, and may be fabricated out of thin air with little or no basis in fact or reality. In other words, an opinion generally represents how someone feels, and may not reflect the truth.

In 2019, a sitting member of the United States Congress said that the world is going to end in 12 years if we don't address climate change. [21] The article cited a 1526-page climate assessment report [22] that revealed no information about the world ending. When questioned, the congresswoman characterized the statement as a rhetorical call to action. Therefore, her statement reflected how she felt. This was the congresswoman's opinion.

Actions based on feelings can create catastrophes around the world that could potentially be much more consequential than yelling "FIRE" in a crowded theater. How many people read headlines, do not read entire articles, do not examine references, do not see retractions, and believe the original headline?

One of many such statements made during the last 100 years was the 1972 prediction that "U.S. oil supplies will last only 20 years. Foreign supplies will last 40 or 50 years but are increasingly dependent upon world politics." [23] Statements predicting the end of oil and gas supplies come and go. When the world's oil was supposed to run out in the early 2020s, the major producers were pumping more oil than when the prediction was made 50 years prior. What was the scientific basis for these claims?

Often, opinions masquerade as facts. For example, someone might say that the air and water in the United States are dirtier now than they were 50 or 100 years ago, when billowing smoke spewed from factories and raw sewage was dumped directly into rivers and lakes. Firefighters were needed to battle flames when the Cuyahoga River in Ohio burned in the early 1950s. Yes, you read that right. The river was on fire.

It is quite interesting that people often come to **diametrically opposite opinions** even when they may agree on the same facts. A person's feelings, beliefs, perspective, understanding, and desires can play in forming an opinion. In other words, two people can start with the same facts that each person emphasizes or downplays differently to reach a different opinion.

For example, two people may examine a graph depicting the global population. One person observes the rapid population increase during the 2000s and concludes that the population will increase until the Earth does not have the means to support everyone. The other person observes the same rapid increase, but calculates recent growth based on the data on the graph and discovers that population growth is slowing. Additional analysis of the population in future decades reveals that the population will stabilize. Both people started with the same information, but their opinions are radically different because the second person utilized more facts.

Regardless of what is being analyzed, it is extremely important to separate fact from opinion. Often, it is helpful to stop, and ask yourself if what you heard or read is a fact or an opinion. Analysis based on facts generally has a better chance of effectively addressing issues by producing superior solutions that are based on reality, as compared to analysis based on opinion.

A reasoned approach will be developed in subsequent chapters to put underlying technologies, societal evolution, technical capabilities, and business viability into perspective. Facts, data, and results based on accepted research and credible evidence of flawed data will be analyzed in context to make effective decisions.

Chapter 3: Evolution of Daily Life

The good ole days weren't always good, and tomorrow ain't as bad as it seems. --- Billy Joel

Many parents send their children off to summer camp for a few weeks to expose them to different places, people, and activities. Other parents just want to take a break. My parents sang the virtues of making new friends and having fun, but it would be reasonable to suspect that they were also happy to have me out of the apartment for a while.

Upon arrival, I was placed in a group with about eight other boys who were also 8 years old. We all slept together in a room that occupied half of a cabin, ate at the same assigned table in the cafeteria, and swam in the lake twice a day (weather permitting). My group shared a latrine located in a nearby building with three or four other cabins. I distinctly remember that the latrine was smelly and that the liquid level rose every day until it almost overflowed through the toilet holes. Then a truck would come, connect hoses, and remove the accumulated liquid. I vaguely recall there being a sink in the latrine and that brushing my teeth was an occasional affair. There were no showers. Instead, we were given soap to clean ourselves in the lake just before going home.

This was my normal as an 8-year-old at summer camp. While seemingly primitive by today's standards, the experience did give me a first-hand experience of how the average person lived not that long ago. Rest assured that I had a great time at camp and was not traumatized. However, I do remember what it was like to *rough it*.

The effect of my living conditions on the environment was negligible and essentially limited to my food consumption and the shared use of two low-wattage light bulbs that were on for about two hours each evening. Life in the United States in the mid-2020s is quite different, with heat, air conditioning, cars, airplanes, restaurants, shopping, and the like, having a much greater effect.

---xxx---

Spoiler Alert: This chapter provides a glimpse of what life was like before humans burned significant amounts of fossil fuels.

Everyday life for the average person was not always as easy as it is today. Many people complain about their lot and do not appreciate what they do have. Examining the evolution of various aspects of our lives will help put our day-to-day lives into historical perspective and guide us regarding the suitability of potential actions related to the environment.

An attempt could be made to describe daily life as it was hundreds and thousands of years ago. However, the details are not necessary. Instead, random discoveries and inventions relating to various aspects of daily life are presented herein to provide an appreciation of what life was like before that discovery or invention. The dates presented are intentionally vague because the overall adaptation of a discovery or invention often occurs decades, and sometimes centuries, after the initial discovery or invention. Further, these dates should impart a sense of appreciation for how profoundly the world and people's enjoyment of that world have changed in a relatively short period of time.

Water, Food, Sewage, and Heat

Water is essential to human life and is more important than food. Humans can live without food for weeks but cannot live without water for more than a few days.

Prior to 1800, digging a well near one's house was often an option to get water, but someone still had to fill water buckets, either manually or using a hand-operated pump, and then carry the heavy buckets to where the water was needed. You might not have to pump if you are one of the fortunate few to tap into a natural spring. People who did not have access to a well or natural spring generally lived near a river or body of fresh water and carried unfiltered unpurified water to their house. Depending on the location and means available, it could take a long time to make the round trip to transport a relatively small amount of water.

Water systems were devised in ancient and modern times to provide water to centralized fountains in cities. Nonetheless, someone had to bring the water from the fountain to the house. In short, providing water was a necessary task that was usually labor-intensive and time-consuming. In contrast, a person can now turn on a faucet and get the desired amount of water at the desired temperature.

Not too many years ago, food might have been stored in a root cellar, which is an underground room whose temperature remains cool throughout the year. Refrigerators for home use were invented in the early 1900s and took decades to reach the average person. Therefore, the convenience of going to the kitchen, opening the refrigerator, and getting some ice is a relatively recent luxury. Similar recent luxuries include machines that make food preparation faster and easier, such as blenders, food processors, bread machines, and the like.

Hearths or wood-burning stoves were often used for cooking and heating, and someone had to gather firewood or get coal and bring it to the house. In addition, someone had to dispose of the ash that remained in the stove after burning the fuel, and the smoke emitted from these fuels adversely affected people's health. These conditions exist in 2023, as suggested in the following excerpt.

> India is blessed with abundant sunlight ... Solar energy based decentralized and distributed applications have benefited those in rural areas by meeting their requirements for lighting, cooking etc. in an environmentally friendly manner. The social and economic benefits include:
> - Eliminates drudgery and health risk among rural women and girls who had to previously collect firewood and toil in smoky kitchens.
> - Generates employment at the village level.
> - Improves standard of living and creates opportunities for economic growth. [24]

Natural gas stoves, electric stoves, microwave ovens, and other cooking appliances that became available to the average household starting in the mid-1900s are relatively recent luxuries that allow us to cook by turning knobs and pushing buttons, as are natural gas and oil furnaces for space heating.

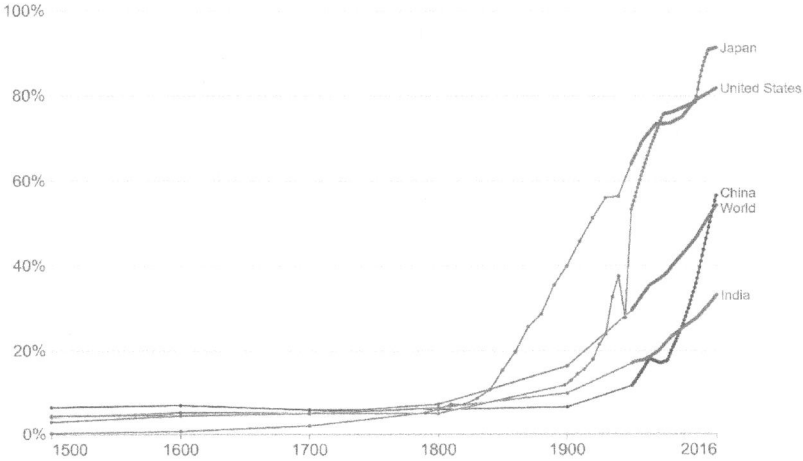

Figure 3-1 Urbanization over the past 500 years, 1500 to 2016 [25]

Obtaining food is another activity that has recently changed dramatically. Figure 3-1 shows that only 16.4 percent of the global population lived in urban areas in 1900. Therefore, most people lived in rural areas and grew much of their own food. Extra food or homemade goods could be sold or bartered to obtain goods and services from others that they themselves could not make or perform, respectively. The variety of foods available was usually limited to what could be produced locally and often did not constitute a healthy diet. The quantity of food available was similarly limited by local food production, so there were times when people went hungry.

Compare this with the recent luxury of going to a nearby bodega, supermarket, or wholesale warehouse, and being able to choose from a selection of thousands of products that the purchasers have no idea how to make.

Thousands upon thousands of restaurants have opened around the world that will not only prepare a hot meal for you but also clean the dishes! Delivery services will even deliver hot meals from your favorite restaurant to your door within minutes. Food services are also available where you can select the meals you want to eat for a week, and they will deliver them to your door and sometimes even

31

put them in the refrigerator for you. This is particularly helpful for people who do not want to cook or do not have time to cook. These luxuries were not commonly available to the average person until recently. In summary, these food options are luxuries when compared to those available not that many years ago.

In some ancient cities, water was diverted to flow under rows of toilets (located in a building) to sweep human waste downstream. Centuries later, the Great Horse Manure Crisis of 1894 was described in this article.

By the late 1800s, large cities all around the world were "drowning in horse manure". For these cities to function, they were dependent on thousands of horses for the transport of both people and goods.

In 1900, there were over 11,000 hansom cabs on the streets of London alone. There were also several thousand horse-drawn buses, each needing 12 horses per day, making a staggering total of over 50,000 horses transporting people around the city each day.

To add to this, there were yet more horse-drawn carts and drays delivering goods around what was then the largest city in the world.

This huge number of horses created major problems. The main concern was the large amount of manure left behind on the streets. On average a horse will produce between 15 and 35 pounds of manure per day, so you can imagine the sheer scale of the problem. The manure on London's streets also attracted huge numbers of flies which then spread typhoid fever and other diseases.

Each horse also produced around 2 pints of urine per day and to make things worse, the average life expectancy for a working horse was only around 3 years. Horse carcasses therefore also had to be removed from the streets. The bodies were often left to putrefy so the corpses could be more easily sawn into pieces for removal.

The streets of London were beginning to poison its people.

But this wasn't just a British crisis: New York had a population of 100,000 horses producing around 2.5m pounds of manure a day.

This problem came to a head when in 1894, The Times newspaper predicted... "In 50 years, every street in London will be buried under nine feet of manure."

This became known as the 'Great Horse Manure Crisis of 1894'.

The terrible situation was debated in 1898 at the world's first international urban planning conference in New York, but no solution could be found. It seemed urban civilisation was doomed.

However, necessity is the mother of invention, and the invention in this case was that of motor transport. Henry Ford came up with a process of building motor cars at affordable prices. Electric trams and motor buses appeared on the streets, replacing the horse-drawn buses.

By 1912, this seemingly insurmountable problem had been resolved; in cities all around the globe, horses had been replaced and now motorised vehicles were the main source of transport and carriage.

Even today, in the face of a problem with no apparent solution, people often quote 'The Great Horse Manure Crisis of 1894', urging people not to despair, something will turn up! [26]

Outhouses and latrines were in common use into the 1950s in rural areas. They could be smelly, hot, or cold in the daytime, and inconvenient and unsafe to visit at night. People rarely took baths but rather cleaned themselves with sponges using water from a washbasin in the house. Naturally, someone had to fetch the water.

Indoor plumbing (commonly called bathrooms or water closets) is the convergence of water distribution systems, sewage drainage systems, and (optional) water treatment systems. Bathrooms commonly include a toilet, sink, and a shower or bathtub. They provided the luxury of attending to one's personal needs without having to brave the outdoor elements. An ample supply of water enabled people to shower every day plus attend to personal hygiene.

Personal Hygiene and Health Care

The first American patent for a toothbrush was obtained in the 1850s, but it took until after World War II for Americans to develop the habit of brushing regularly. [27] Americans have since adopted the routine of getting periodic preventative dental examinations and x-rays.

Religious handwashing rituals have been around for thousands of years in Islamic, Jewish, and other cultures. However, the notion of disease spreading by hand has only been part of the medical belief system since around 1850. [28] Ignaz Semmelweis, a Hungarian doctor working in Vienna General Hospital, is known as the father of hand hygiene. In 1846, he noticed that the women giving birth in the medical student/doctor-run maternity ward in his hospital were much more likely to develop a fever and die compared to the women giving birth in the adjacent midwife-run maternity ward. [29] Sadly, the hand hygiene practices promoted by Semmelweis and (Florence) Nightingale were not widely adopted. In general, handwashing promotion stood still for over a century. It was not until the 1980s, when a string of foodborne outbreaks and healthcare-associated infections led to public concern that the United States Centers for Disease Control and Prevention identified hand hygiene as an important way to prevent the spread of infection. [30]

The previous paragraph provides an insight into one aspect of the generally squalid hospital conditions that existed in the 1800s. Treatments that would now be considered grotesque focused on symptomatic relief. People had good reasons to be careful to avoid accidents and keep from getting sick. People used home remedies to avoid the unpleasantness, health risks, and high cost of being treated. Routine annual examinations and periodic visits to specialists are more recent developments.

Florence Nightingale (1820-1910) was an English statistician who founded modern nursing and greatly improved patient survival rates. Other major medical advancements include the development of the smallpox vaccine (early 1800s), the stethoscope (early 1800s), the cholera vaccine (late 1800s), the thermometer (late 1800s), the Penicillin antibiotic (early 1900s), and antipsychotics (mid 1900s).

The importance of washing one's hands before and after eating, and after using the bathroom is a relatively recent development that is now well-established.

Electricity

Several scientists researched magnetic fields caused by electricity in the early 1800s. Only in 1882 did Edison help form the Edison Electric Illuminating Company of New York, which brought electric light to parts of Manhattan. Subsequent progress was slow, and most Americans still lit their homes with gas lights and candles for another fifty years. Only in 1925 did half of all homes in the U.S. have electric power. [31] By 1950, close to 80 percent of U.S. farms had electric service. [32]

Ubiquitous adaptation of electricity and light bulbs enabled many of the inventions and innovations of the 1900s.

Communication

Personal communications were largely person-to-person, a variation of one person telling another person to say something to a third person, written messages, or the mail. The invention and implementation of the telegraph in the mid-1800s enabled messages to travel long distances within minutes instead of via written letters carried by horse-and-carriage systems that could take weeks to be delivered. The telephone, invented in the late 1800s and adopted by the population in the early to mid-1900s, allowed for instantaneous verbal communication over long distances. Since the 1990s, cell phones, the Internet, smart phones, texting, social media, and videoconferencing have come into widespread use, providing the luxury of virtually instantaneous communication with virtually everyone in the world. The Internet also provides the previously unimaginable luxury of having immediate access to a wide range of human knowledge.

In just a few short decades, personal communication evolved from a slow and labor-intensive process to a virtually instantaneous action.

Mass communication has evolved from newspapers and magazines to the first radio broadcast (1920) to the first television broadcast (1928) to the first live television broadcast (1951) to the first cable television broadcast (1948) to the Internet that was implemented in the 1980s. General integration of these inventions into everyday life took years, and full integration is still an ongoing evolutionary process.

Transportation

Transportation systems on land generally evolved from walking to riding animals to the horse-and-buggy to trains (early 1800s) to automobiles and buses (late 1800s) to airplanes (early 1900s) to space travel (1950s). It typically took decades of innovation for these technologies to become available to the average person. Therefore, jumping into your car to drive somewhere on a minute's notice, or hopping on a plane for a weekend getaway on another continent are relatively recent luxuries.

Education

The importance of education cannot be understated as a catalyst for people to improve their standard of living. Education was haphazard before public schools. Many children were excluded based on income, race or ethnicity, gender, geographic location, and other reasons. [33] Starting in approximately 1850, various states in the United States passed laws that made education mandatory for children under their jurisdiction.

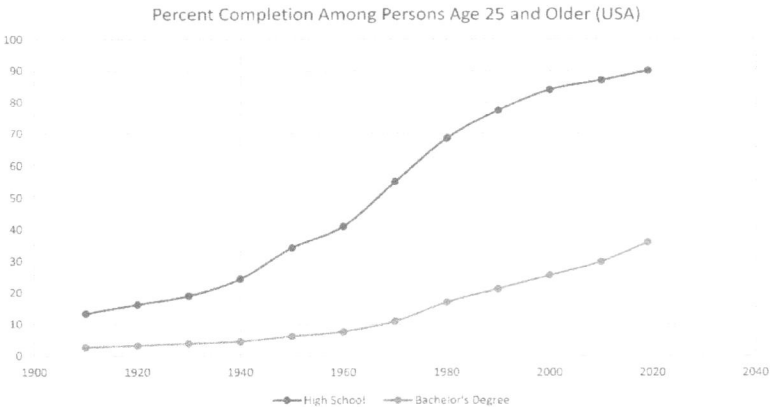

Figure 3-2 Percent Completion Among Persons Ages 25 and Older (USA) [34]

By around 1920, all children in the United States were required to attend school. Figure 3-2 shows that the high school degree completion percentage in the United States steadily improved in the mid-1900s and into the early 2000s. The bachelor's degree percentage likewise increased before accelerating in the mid-1900s.

36

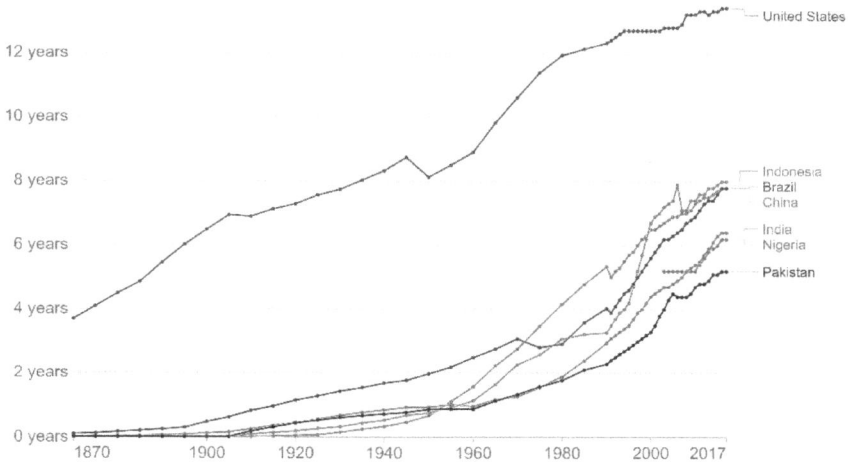

Average years of schooling
Average number of years people aged 25+ participated in formal education.

Source: Lee-Lee (2016); Barro-Lee (2018) and UNDP HDR (2018)
OurWorldInData.org/human-development-index • CC BY
Note: Formal education is primary/ISCED 1 or higher. This does not include years spent repeating grades

Figure 3-3 Average years of schooling [35]

Figure 3-3 shows that the amount of education people received has increased dramatically since the mid-1900s in countries with populations of over 200 million people, which collectively comprise approximately half of the world's population.

Voting Rights

In the late 1700s, voting in the United States was limited to white male landowners. Voting rights were expanded in different localities in the 1800s and 1900s. Constitutional amendments included the 15th Amendment for African American men (1870), the 19th Amendment for women (1920), the 24th Amendment eliminating polling taxes (1964), and the 26th Amendment for 18-year-old voters (1971). Despite these amendments, some states used literacy tests and other barriers, to make it harder for certain groups to vote.

Prejudice, Discrimination, and Slavery

For the purposes of this discussion, *prejudice* is a preconceived, usually negative, judgment or opinion formed without just grounds or sufficient knowledge. *Discrimination* is the prejudiced outlook, action, or treatment of different categories of people that can take different forms, such as by ethnicity, race, age, disability, national origin, and appearance. *Slavery* is a situation or practice in which people are entrapped and exploited, such as by ownership and indebtedness.

Prejudice is rampant and can apply to mundane situations. Everyone does it to some degree. It cannot be stopped, but it can be controlled. It is also normal. For example, would you take a leisurely walk with your infant child in Central Park, New York City, at 3 a.m. in the morning? No! Prejudice, because you have prejudged that people there at that time might do you harm. Do you remember the 1993 film 'Cool Runnings' where a sprinter disqualified from the Olympic Games started the first Jamaican bobsled team? People thought that it was a joke. Prejudice, because countries in tropical climates do not compete in such events. The film was inspiring, but the Jamaicans did not win. If all of the players on Brazilian World Cup soccer team were replaced, would there be any doubt that Brazil would qualify for the next World Cup? Prejudice, because people think that Brazil has so much depth that any Brazilian team would qualify. Not qualifying would be considered a national disgrace. In contrast, not winning the World Cup tournament outright would only be embarrassing.

Discrimination has been around in various forms for a long time, but it has been seen as normal rather than negative. [36] For example, to this day, there are traditional 'Japanese Only' restaurants that only serve Japanese people, complete with signs so stating. This has been part of Japanese culture for centuries and is not considered discrimination by the Japanese, because they eat in them all the time. [37] Non-Japanese people would probably disagree with this practice and attitude. Racially homogenous countries, like those in Africa, discriminate based on ethnicity, tribe, religion, lineage, and race. [38] For example, in Liberia, citizenship is restricted to those who are black. [39]

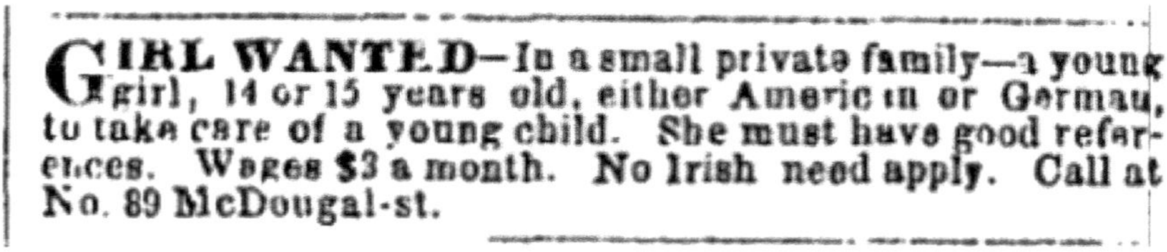

> GIRL WANTED—In a small private family—a young girl, 14 or 15 years old, either American or German, to take care of a young child. She must have good references. Wages $3 a month. No Irish need apply. Call at No. 89 McDougal-st.

Figure 3-4 Help Wanted Advertisement (1854) [40]

Employment opportunities and education were denied to many who were different, as illustrated in Figure 3-4. In the 1800s, people already established in the United States openly discriminated against newly arrived Irish, Italian, and Jewish immigrants. [41]

However, discrimination was in no way limited to the United States. People in every nation discriminate, and it is often legal. [42] Countries with homogeneous populations can be quite discriminatory. Examples include the aforementioned 'Japanese Only' restaurants in a country that is 98 percent ethnically Japanese, and discrimination against ethnic groups in China other than the Han, which comprise 92 percent of the Chinese population.

Slavery has been around since biblical times and has since taken different forms.

> Now these are the judgments which thou shalt set before them. If thou buy an Hebrew servant, six years he shall serve: and in the seventh he shall go out free for nothing. If he came in by himself, he shall go out by himself: if he were married, then his wife shall go out with him. If his master have given him a wife, and she have born him sons or daughters; the wife and her children shall be her master's, and he shall go out by himself. [43]

Chattel slavery is the total ownership of a person, which was common in the southern United States until the Confederacy was defeated in the Civil War, which claimed approximately 620,000 lives, the overwhelming majority of which were white. This is approximately equal to the total of American fatalities in the Revolutionary War, the War of 1812, the Mexican War, the Spanish American War, World War I, World War II, and the Korean War, combined. [44]

Chattel slavery was not unique to the United States. It was present among Europeans, Arabs, indigenous American Indians, and others over the centuries. [45] For example, historians estimate that between 10 million and 20 million people were enslaved by Arab slave traders between 650 and 1900. Boys between the ages of 8 and 12 were castrated to prevent them from reproducing, and about six of every 10 boys bled to death during the procedure. [46]

To escape poverty in their home country, some people indentured themselves to work for a period of years to pay for the cost of transportation to the United States. This indebtedness was a form of slavery, where many worked dangerous jobs and did not live long enough to repay the debt to regain their freedom. [47] Others worked in brothels, factories, mines, and on farms.

Perspective

For much of the 20th century, most people lived in rural areas and were largely self-sufficient. This meant building things that are not available locally, fixing whatever breaks, growing food, and raising animals. These are activities that need to be performed seven days a week, with very few days off for rest or relaxation. Manufactured goods could be acquired by using money earned from selling extra food or by bartering with others for their food and services.

The integration of personal communications, mass communications, and transportation systems has enabled the faster and more efficient movement of goods and services. The ability to order a part on the Internet from a company in California and have it delivered in New York by 10 o'clock the next morning is a luxury that has only recently become an option.

The number of people working in offices and factories increased during the 20th century as people relocated to urban areas. The integration of the computer into the workplace allowed people to work at virtually any time from almost anywhere in the world.

People in urban areas have been watching plays and listening to musical performances since ancient times. More recently, everyday entertainment has evolved from family members playing music together to watching a movie to listening to the radio to watching television to playing video games to streaming films to attending video meetings with friends and family. Stated differently, there were many fewer entertainment options available just a few decades ago.

Not that long ago, the day-to-day life of an average person included many activities that we now consider mundane. However, they were necessary for survival and consumed a considerable amount of time and effort because most people lived in rural environments and had to *do or do without*. Their lives were focused on survival. They did what they needed to do. Recent inventions and innovations have made these activities considerably easier or simply not necessary. For example, growing and canning food for later use has evolved from a mandatory activity for survival into a hobby for those who enjoy growing and canning food.

The world has evolved such that many people can spend much of their time focused on more satisfying and productive activities. Now people can play professional baseball, wait tables, build bridges, drive a taxi, or practice a profession in which they are qualified (and hopefully enjoy) during the week, and visit a museum and see a play on the weekend instead of fetching water and firewood, tending to the animals, and eating mush every day like the average person did in 1800.

Think about that for a minute. Much has changed in such a short time, and this evolution continues as more and more people raise their standard of living.

Chapter 4: Transition from a 'Life of Death' to a 'Life of Life'

When you come to a fork in the road, take it! --- Yogi Berra

I distinctly remember concluding that anything and everything could be done in one day. The days were just so long, and never seemed to end. By the way, I was only six years old at the time, and I really did not have a lot to do. Having aged since, I recognize that my horizons were quite limited.

---xxx---

Spoiler Alert: This chapter presents historical data that can be used to project whether the world population continues to grow exponentially, stabilizes, or shrinks, and when it will do so. This information will be helpful to determine the actions, if any, that should be taken and when they should be implemented.

Actions taken, or not taken, in the name of **climate change** can influence virtually every aspect of life to some extent. Selecting the proper actions requires not only a cursory understanding of multiple technical disciplines but also a general appreciation of how everyday life has evolved over time and why. We will discover that the average person's life prior to around 1900 was not that pleasant, and moving backwards to relive those times is not a good idea. The positive side of this discovery is that significant progress has been made to eliminate human suffering, which has been the human condition for thousands of years.

Malthusian Theory

Thomas Malthus (1766-1834) was an English economist and clergyman who theorized that infinite human hopes for social happiness were in vain because unchecked geometric growth in population will always tend to outrun its arithmetic growth in resources. Therefore, the population will only expand to its limit of subsistence, resulting in unchecked population growth that would cause widespread poverty and degradation. Malthus said that the population cannot grow beyond that for which its resources can provide support, because if the population does grow excessively, people will starve for lack of resources, and the population will decrease. Therefore, "an obvious truth (is that the) … population must always be kept down to the level of the means of subsistence." [48] Malthus stipulated that excessive population growth could be checked by vice (including the commission of war), misery (including famine or want of food and ill health), and moral restraint, suggesting that the betterment of mankind is impossible without stern limits on reproduction, recommending late marriage and sexual abstinence as methods of birth control. [49] [50]

Malthus essentially described what is now known as the Malthusian Trap, where resources are limited, and humans are forever trapped in lives of subsistence.

> Population, when unchecked, increases in a geometrical ratio. Subsistence increases only in an arithmetical ratio. A slight acquaintance with numbers will shew the immensity of the first power in comparison of the second.
>
> By that law of our nature which makes food necessary to the life of man, the effects of these two unequal powers must be kept equal.
>
> This implies a strong and constantly operating check on population from the difficulty of subsistence. This difficulty must fall somewhere and must necessarily be severely felt by a large portion of mankind. [51]
>
> But though the rich by unfair combinations contribute frequently to prolong a season of distress among the poor, yet no possible form of society could prevent the almost constant action of misery upon a great part of mankind, if in a state of inequality, and upon all, if all were equal.
>
> The theory on which the truth of this position depends appears to me so extremely clear that I feel at a loss to conjecture what part of it can be denied.
>
> That population cannot increase without the means of subsistence is a proposition so evident that it needs no illustration. [52]

Malthus appears to identify uncontrollable factors that limit population growth and suggests measures to proactively control the population to reduce human suffering in the future.

Malthus wrote his essay in 1798 about the world in which he lived with the benefit of having access to centuries of history. The general condition of the human population was one of subsistence that was focused on survival. Wars, diseases, and famines that regularly limited population growth and, in some cases, reduced the population substantially, can be readily chronicled during the preceding centuries.

Wars with at least one million deaths accounted for well over 180 million deaths during the 1800 years preceding Malthus' essay in 1798.[53]

- Jewish-Roman Wars in the Middle East from 66 to 136 (1.27 to 2 million deaths)
- Three Kingdoms War in China from 184 to 280 (36-40 million deaths)
- Yellow Turban Rebellion in China from 184 to 205 (3-7 million deaths)
- Arab-Byzantine Wars from 629 to 1050 (2+ million deaths)
- Reconquista in Spain from 711 to 1492 (7 million deaths)
- An Lushan Rebellion in China from 755 to 763 (13 to 36 million deaths)
- Crusades in the Middle East from 1095 to 1291 (1 to 3 million deaths)
- Mongol invasions and conquests in Eurasia from 1206 to 1368 (30 to 40 million deaths)
- Hundred Years' War in Western Europe from 1337 to 1453 (2.3 to 3.5 million deaths)
- Conquests of Timur in Eurasia from 1370 to 1405 (8 to 20 million deaths)
- Spanish conquest of the Aztec Empire in Mexico from 1519 to 1632 (24.3+ million deaths)
- Spanish conquest of the Yucatan in North America from 1519 to 1595 (1.46+ million deaths)
- Spanish conquest of the Incan Empire in Peru from 1533 to 1572 (8.4+ million deaths)
- French Wars of Religion in France from 1562 to 1598 (2 to 4 million deaths)
- Japanese invasion of Korea from 1592 to 1598 (1+ million deaths)

- Transition from Ming to Qing in China from 1616 to 1683 (25+ million deaths)
- Thirty Years War in Europe from 1635 to 1659 (4 to 12 million deaths)
- Deluge in Poland from 1648 to 1657 (3 million deaths)
- Mughal-Maratha Wars in India-Bangladesh from 1658 to 1707 (5+ million deaths)
- Tay-Son Rebellion in Southeast Asia from 1771 to 1802 (1.2 to 2 million deaths)

Similarly, epidemics with at least one million deaths accounted for well over 110 million deaths during the 1800 years preceding Malthus' essay in 1798.[54]

- Antonine Plague in the Roman Empire from 168 to 180 (5-10 million deaths)
- Plague of Justinian primarily in Europe from 541 to 549 (15 to 100 million deaths)
- Japanese Smallpox Epidemic in Japan from 735 to 737 (2 million deaths)
- Black Death in Europe from 1346 to 1353 (75 to 200 million deaths)
- 1520 Mexican smallpox epidemic in Mexico from 1519 to 1520 (5 to 8 million deaths)
- Cocoliztli epidemic in Mexico from 1545 to 1548 (5 to 15 million deaths)
- Cocoliztli epidemic of 1576 in Mexico from 1576 to 1580 (2 to 2.5 million deaths)
- Italian Plague in Italy from 1629 to 1631 (1 million deaths)
- Naples Plague in Italy from 1556 to 1558 (1.25 million deaths)
- Persian Plague in Persia from 1772 to 1773 (2 million deaths)

The data for these wars, diseases, and famines are estimates of the deaths that occurred and do not reflect how many people were injured but not killed. In addition, numerous other wars, epidemics, and famines occurred that were smaller in size, but collectively added significantly to the death toll.

To put this data into perspective, note that the estimated population of the world in the year 0 and in 1800 per Figure 4-1 is approximately 190 million and one billion people, respectively. Therefore, many of these events significantly reduced the human population at the time. Some of the early epidemics are estimated to have reduced local populations by 25 percent to as much as 80 percent.

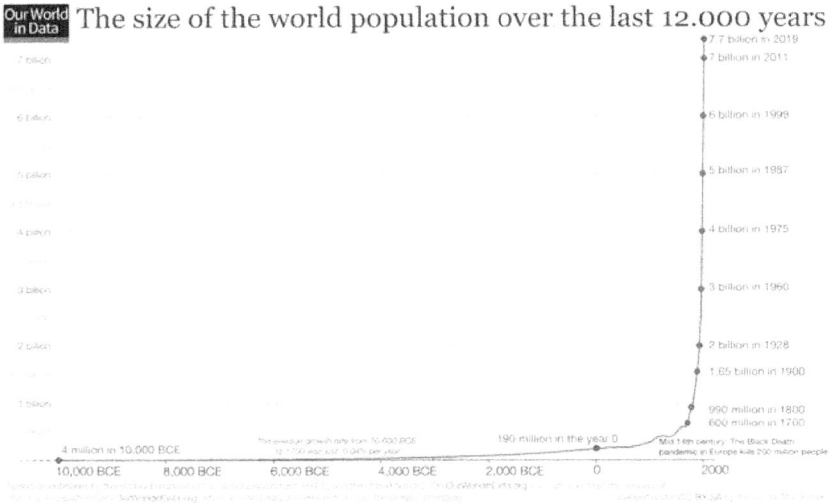

Figure 4-1 The size of the world population over the last 12,000 years [55]

Such was the state of the world when Malthus wrote his essay. However, Malthus had no way to predict that an evolution from this state to another was in its infancy, and that the transition would probably take three centuries to complete.

Child Mortality

Child mortality is the probability of a child dying between a live birth and the age of five years. In 1800, the global child mortality rate was approximately 43.3 percent, as shown in Figure 4-2. Related information in Figure 4-3 suggests that the global child mortality rate was somewhat higher during the preceding 1800 years. Therefore, it is reasonable to estimate that the child mortality rate between the years 1 and 1800 was approximately 45 percent.

Global child mortality

Share of the world population dying and surviving the first 5 years of life.

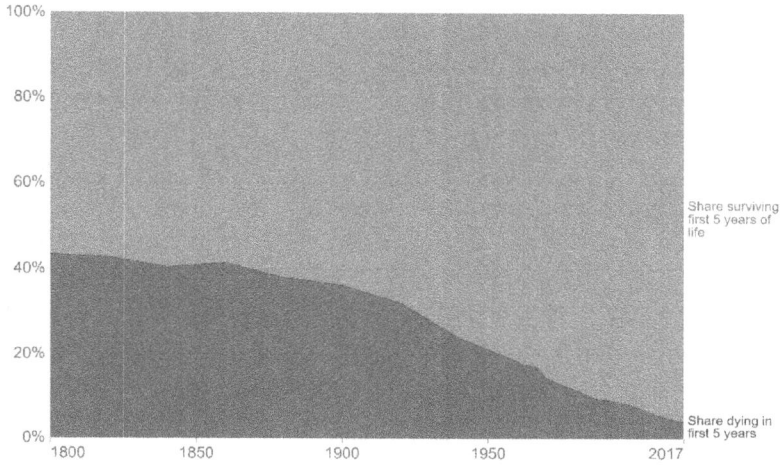

Figure 4-2 Global child mortality [56]

Mortality rates of children over the last two millennia

● The youth mortality rate: the share who died as infants or children (younger than 15).
● The infant mortality rate: the share who died in their first year of life.

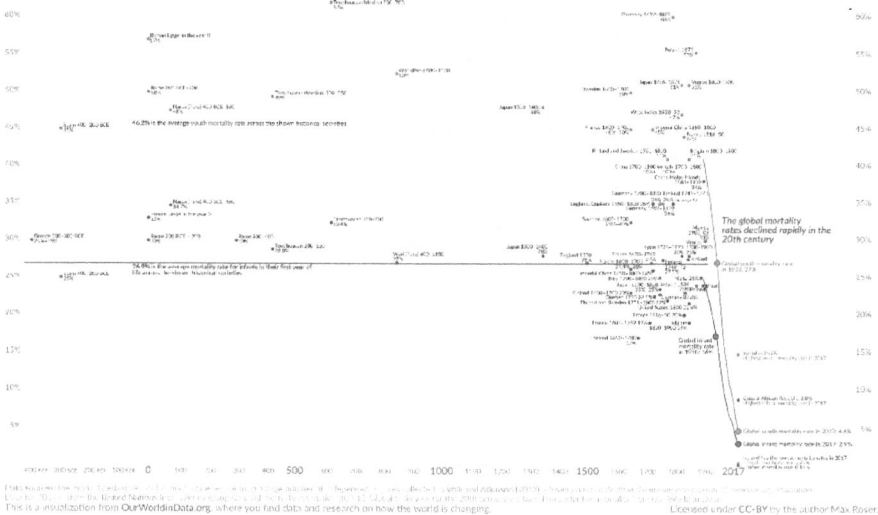

Figure 4-3 Mortality rates of children over the last two millennia [57]

Read the previous paragraph again slowly, and then think about it in human terms for a minute. The data shows that only slightly more than half of all children born live to reach adulthood.

In other words, **almost half of all children died before reaching the age of five**.

It is instructive to understand that utilizing other data can often provide more insight, and that not using all applicable data can limit understanding. For example, life expectancy is an estimate of the average age at which a group of people will be when they die. Life expectancy was approximately 30 years for people born in 1800. [58] However, the child mortality data presented above shows that nearly half of the population born in 1800 died before their fifth birthday. For life expectancy in 1800 to be 30 years, a little more than half of the population had to be approximately 60 years old when they died. Utilizing both child mortality and life expectancy provides additional insight into life in 1800. Considering only life expectancy could erroneously lead one to believe that people born in 1800 only lived to the age of 30.

The response to this grim reality was for parents to have more children to ensure that some of their children survived and had children. More self-serving, parents needed children to help with day-to-day work to survive and, later, to provide care for the parents as they aged. Under these circumstances, it should not be surprising that Figure 4-4 shows that women averaged almost six babies each in 1800, because this was necessary to stabilize the population.

AVERAGE NUMBER OF BABIES PER WOMAN FROM 1800 TO TODAY

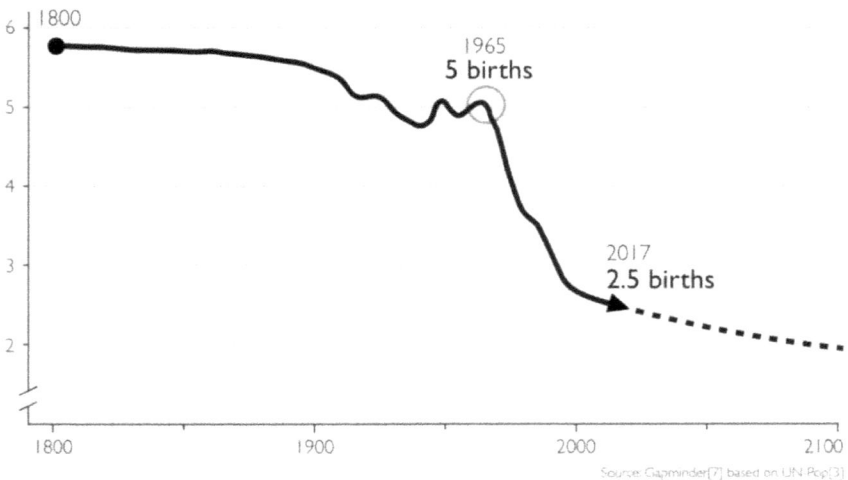

Figure 4-4 Average Number of Babies per Woman from 1800 to Today [59]

48

Whereas carrying a child changes the body, giving birth is a traumatic shock to the body. Limited data in Figure 4-5 indicates that mothers were at risk of death at a rate of approximately 900 per 100,000 live births in Sweden from 1751 until 1820, which is almost 1 percent of live births. The maternal mortality rate was probably higher than one percent prior to 1800, although little data is available.

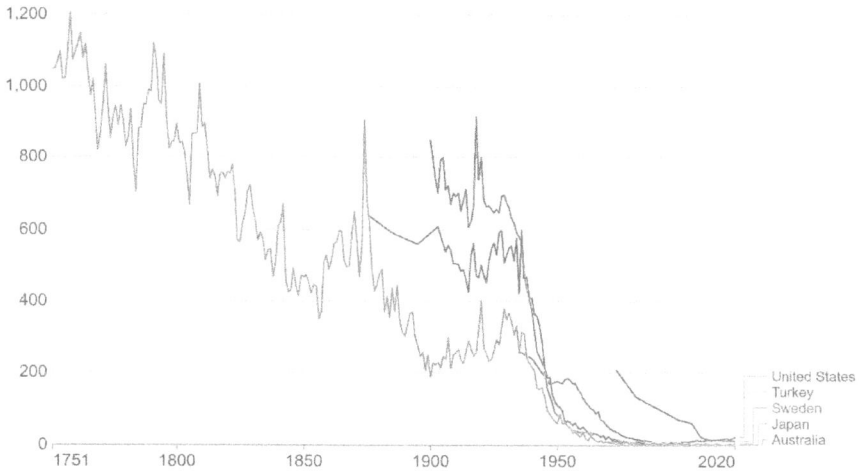

Maternal mortality ratio, 1751 to 2020

The maternal mortality ratio is the number of women who die from pregnancy-related causes while pregnant or within 42 days of pregnancy termination per 100,000 live births.

Source: Gapminder (2010); WHO (2019), OECD (2022)

OurWorldInData.org/maternal-mortality • CC BY

Figure 4-5 Maternity mortality ratio, 1751 to 2020 [60]

Nonetheless, Figure 4-1 shows that the world population is estimated to have increased approximately fivefold between the years 1 and 1800.

In summary, the world prior to approximately 1800 was a world that was in equilibrium with death. This world was not pleasant for the average person, who was in a constant struggle to survive famines, health issues, and wars that could drag on for decades and, sometimes, centuries.

Urbanization and Unions

The Industrial Revolution that began in England was significantly advanced by James Watts' improvements to the steam engine in the

1760s along with the presence of a favorable business environment (capitalism), among other factors.

Labor was needed to operate the machines, so people moved from their familiar rural environment to cities and towns near factories. Previous advances had improved agricultural efficiency so much that the remaining farmers could support people in the cities who no longer produced their own food.

Figure 3-1 shows that the percentage of people who lived in urban areas worldwide was essentially stable until approximately 1800. Between 1800 and 1900, the urban population approximately doubled and then more than doubled again by 2000.

Note that the United States urbanized quickly in the 1800s and 1900s as workers flocked to cities pursuing better work opportunities despite often living in cramped and squalid conditions. Workers were generally not able to feed themselves without a paycheck, so most became beholden to the whims of factory owners, who could hire newly arrived workers as replacements. The labor movement in the United States, formed to protect the interests of workers, grew in strength during the 1800s. However, early union representation was by no means perfect because "union members in the skilled trades remained overwhelmingly native-born White Protestant males throughout the 19th century" and "were reluctant to organize unskilled Irish and Italian immigrants, and also excluded women and Black workers," [61] prompting some groups to form their own unions.

Over the decades, unions were able to secure significant improvements, including better and safer working conditions, health benefits, retirement benefits, paid holidays and vacation, sick days, maternity leave, and a minimum wage, some of which were enacted into law. Conversely, after 200 years of industrialization, employers have generally come to recognize that abusing employees is not a good business practice, and many companies bend over backwards to retain good employees and ensure that their needs are satisfied.

Union membership has dropped from 20.1 percent of wage and salary workers in the first year reported (1983) to only 10.1 percent in 2023. Aside from illustrating the diminishing influence of unions, close to half of current union members work in the public sector. [62] Interestingly, the unions that were organized to protect workers against actual abuse by ruthless employers are now representing an almost equal number of government workers as private sector workers. This begs explanations of whether the government sector has abused workers in the past, whether the government is currently abusing workers, and why government employees need protection at all given that government workers appear to be treated well and sometimes better than non-governmental workers.

Child Mortality (Revisited)

Figure 4-6 reveals that global child mortality in the United States exhibited a downward trend in the 1800s, followed by a more rapid decline in the 1900s. Child mortality was approximately 46, 32, 4, and 0.7 percent in 1800, 1870, 1950, and 2020, respectively.

Child mortality rate (under five years old) in the United States, from 1800 to 2020*

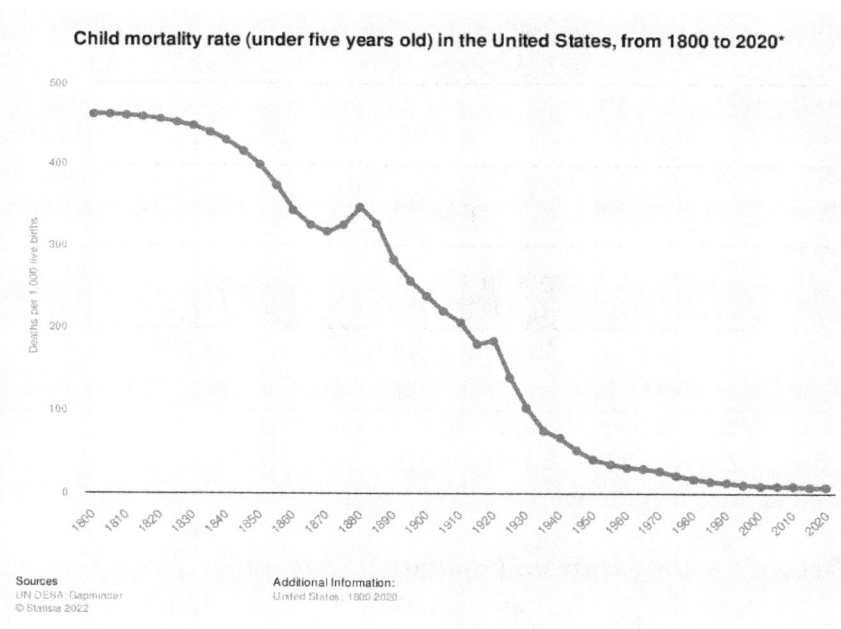

Sources
UN DESA; Gapminder
© Statista 2022

Additional Information:
United States; 1800-2020

Figure 4-6 Child mortality rate (under five years old) in the United States, from 1800 to 2020 [63]

After hundreds of years of relative stability, the drop in child mortality from approximately 45 percent to 0.7 percent in approximately 200 years is not only dramatic, but fantastic news in human terms. No longer were almost half of all babies being born destined to die before they reach their fifth birthday, so most parents

51

no longer have to bear the pain of losing their children. This extremely positive change is almost never discussed or even recognized, let alone analyzed.

This dramatic drop in child mortality also suggests that something happened around the year 1800 that precipitated the fall. Maybe the French Revolution or the United States becoming an independent nation in the late 1700s caused the drop. But those were political events that did not affect babies. Maybe the Industrial Revolution caused people to change, but that started in England. Maybe urbanization resulted in better health care, but hospitals were appalling in the 1800s. Maybe urbanization resulted in more healthcare research, from which life-saving vaccines, drugs, and surgeries were discovered, but that was generally not the case in the 1800s. Maybe urbanization resulted in better living conditions, but housing in cities was relatively poor for newcomers. Maybe people took better care of themselves, but the need for personal hygiene had not yet evolved fully. Maybe people obtained a better education, but most people had to work to survive and did not have a lot of time for school. Maybe urbanization provides better access to healthcare. Maybe...

All the suggested items in the previous paragraph exerted some amount of influence on the drop. An investigation into the dominant causes of the dramatic drop in child mortality might be interesting, but it does not matter. The drop happened for whatever reason and caused demographic changes that started around 1800 and are still ongoing in the 2020s. However, how these changes affected and continue to affect the world, are important.

Effect of Fertility Rate on Population Growth

In Figure 4-6, child mortality dropped from approximately 46 percent in 1800 to approximately 4 percent in 1950. In Figure 4-4, the birth rate remained at approximately 5.5 babies per woman in the 1800s and then oscillated in the mid-1900s during World War I, World War II, and the Great Depression. Despite the oscillations, the birth rate was relatively stable at approximately 5 babies per woman until around 1965.

In years prior to 1965, women had approximately the same number of babies, but these babies did not die as children, so the parents found themselves with more living children to feed, clothe, and educate. In addition, more women were alive to give birth each year, because fewer babies died, and more babies lived into their reproductive years. The net result was that the population increased from approximately 1 billion people in 1800 to approximately 1.65 billion people in 1900, approximately 3 billion people in 1960, and approximately 6.5 billion people in 2000.

Examined differently, the global population grew by approximately 65 percent during the 100 years of the 1800s, approximately 82 percent during the next 60 years, and approximately 117 percent during the next 40 years. In other words, the population grew exponentially as successively shorter time periods exhibited successively increasing population growth. This observation gives the impression that population growth is accelerating, will continue to accelerate, and will not stop accelerating until, at some point, the Earth will not be able to support all its inhabitants.

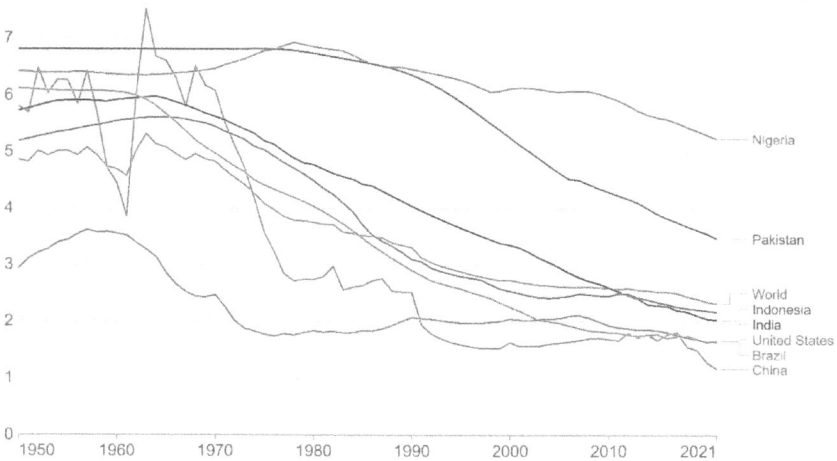

Fertility rate: children per woman

Source: United Nations - Population Division (2022)   OurWorldinData.org/fertility-rate • CC BY
Note: The total fertility rate is the number of children that would be born to a woman if she were to live to the end of her child-bearing years and give birth to children at the current age-specific fertility rates

Figure 4-7 Fertility rate: children per woman [64]

However, Figure 4-7 shows that the average number of babies born to each woman in the world dropped dramatically from its historical value of over 5, to approximately 2.32 children per woman in 2021. Figure 4-7 also shows the global fertility rate, and fertility rates for countries with populations over 200 million people, which collectively comprise approximately half of the world population. They are Nigeria (5.24), Pakistan (3.47), World (2.32), Indonesia (2.18), India (2.03), United States (1.66), Brazil (1.64) and China (1.16). All the trends exhibit a downward trajectory. Of note is that the countries with the largest populations (India, China, and the United States) have birth rates below the 2.10 considered appropriate to maintain the current population.

The populations of these countries can be expected to fall if they stay in their current trajectories. The population decline in at least one country has already started, as illustrated by China's National Bureau of Statistics announcement that China's population fell to 1.411 billion people in 2022, which is 850,000 fewer people as compared to 2021. [65] It fell another 2.08 million in 2023. [66]

Population Growth

The near-vertical line on the right side of Figure 4-1 shows that the population of the world is increasing dramatically, often referred to as exponentially. A cursory examination of this near-vertical line on the graph would lead an observer to conclude that the current trend will continue for a long time to come, if not forever. This is the Rosling straight-line instinct at work. Such an occurrence would not be sustainable because the additional people would increase demand for food and goods that the Earth may not be able to provide.

Despite what appears to be exponential population growth in Figure 4-1, the United Nations projects the world population to reach 8.5 billion people in 2030, 9.7 billion in 2050, and 10.4 billion by 2100. Other United Nations projections include Africa having the largest growth rate and Europe losing population. [67] Figure 4-8 attests to the reasonableness of these projections because by 2020, the birth rates in some countries will have trended downward below 2.10 children per woman, except for Africa, which is still appreciably above 2.10.

54

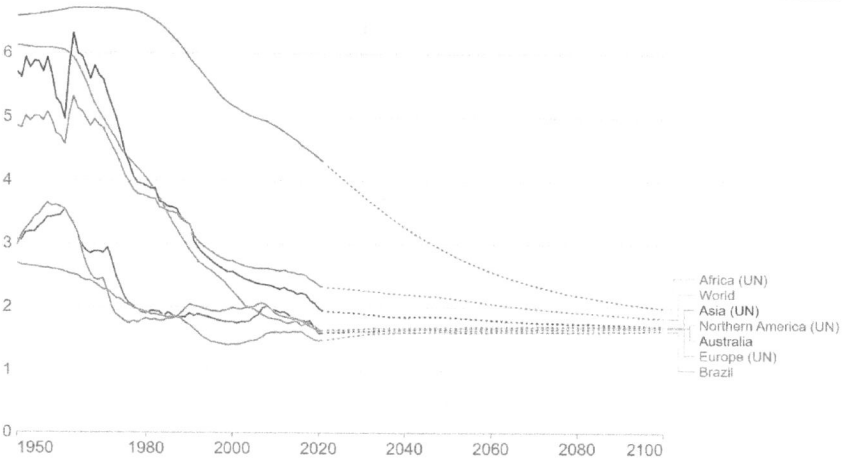

Figure 4-8 Fertility rate: children per woman including UN projections, 1950 to 2100 [68]

The reduced birth rate per woman provides a strong tailwind to achieving a stable world population, whereas increased longevity is a headwind.

The methodology for projecting the world population in the future is complicated, but the concept can be illustrated using the hypothetical numbers in Table 4-1.

Table 4-1 Hypothetical World Population Projection

Age (years)	2020 [69]	2040	2060	2080	2100
80+ (estimate)	0.1	0.2	0.3	0.5	0.8
60-80	0.9	1.8	2.3	2.6	2.3
40-60	1.8	2.3	2.6	2.3	2.6
20-40	2.3	2.6	2.6	2.6	2.6
0-20	2.6	2.6	2.6	2.6	2.6
Total --->	7.7	9.5	10.4	10.7	10.9

The first column of data contains a hypothetical global population distribution by age in 2020. Each successive column assumes an effective birth rate of approximately 2.10 and contains the distribution of people from the previous column when they will be 20 years older. For example, if no one dies, the same 2.6 billion people who were 0-20 years old in 2020 will be 20-40 years old in 2040, 40-60 years old in 2060, 60-80 years old in 2080, and some more than 80 years old in 2100. The 2.6 billion people who were 0-20 years old in 2040 were born to the 2.6 billion people who were 20-40 years old in 2040. The reality of people dying when they are less than 80 years old and the birth rate falling somewhat below 2.10 will reduce these projections and make the numbers closer to those published by the United Nations.

While not rigorous, the general methodology used in this table gives us an understanding of why the global population will grow rapidly until approximately 2060, slow dramatically, and then stabilize.

A closer look at Figure 4-1 showing the size of the world population over the last 12,000 years, indicates that the population grew by 16.7 percent during the 12-year period starting in 1999 (averaging +1.39 percent annually) and by 10 percent during the 8-year period starting in 2011 (averaging +1.25 percent annually). Additional data [70] cites an annual population change of +1.05 percent in 2020. Therefore, actual data shows that the rate at which the global population is increasing, is slowing. This is consistent with the falling birth rates cited above.

Figure 4-9 depicts the actual world population from 1800 to 2020 and the projected world population from 2020 to 2100. [71] Note how the curve bends around 2060, and then stabilizes at approximately 10.5 billion people around the year 2100.

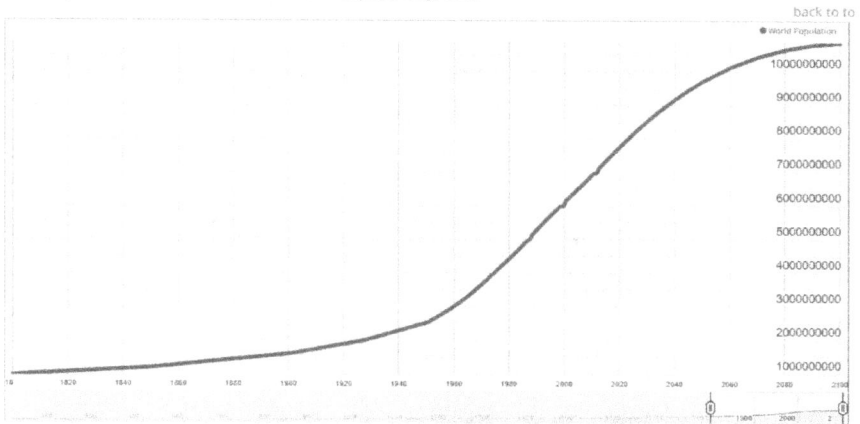

Figure 4-9 World Population: Past, Present, and Future [72]

Significantly, various predictions suggest that China's population could drop from over 1.4 billion people in the early 2020s to as low as 700 to 800 million people by 2100. This projection of a dramatic population reduction is the result of a broad effort from 1970 to 2021 to control China's population, which included the reputedly brutal enforcement of families only being allowed to have one child between 1985 and 2015.

People in China born into larger families prior to 1970 will start to die in significant numbers starting in the 2020s. They will be replaced with fewer younger families that have fewer children. In subsequent decades, more people will die than will be born, and the Chinese population will drop gradually by an estimated 600 million people. As this process unfolds, caring for its elderly with fewer young people will put significant social and economic burdens on China.

The disappearance of this large population will significantly affect the environment in a positive manner, because this phantom population will not consume resources that contribute to emissions. However, the population drop in China will significantly offset population gains elsewhere and help stabilize the global population.

In summary, the global population is amid what appears to be a three-century transition from approximately one billion people in 1800 to ten billion people in 2100. This transition appears to have started due to the convergence of many factors and developments.

57

The preceding paragraphs illustrate the importance of examining multiple pertinent facts when analyzing a given situation or event. In this analysis, a long-term graph of the global population appears to show that exponential population growth will continue. However, when considering other facts and data in the analysis, it appears that the global population is growing at a slower rate and is projected to stabilize. Note that the initial observation of exponential growth based on one piece of information from one graph is downright scary, whereas the determination of reaching a stable population based on more complete information, is more measured and more realistic.

Predictions

Figure 4-7 above shows that the fertility rate for the six largest countries by population has been trending downward for decades and continues to trend downward. This will provide a tailwind to reduce global population growth, and potentially reduce the global population should the global fertility rate of approximately 2.32 in 2021 fall below approximately 2.10. On the other hand, people tend to consume more resources as they improve their standard of living, which will increase carbon dioxide emissions.

It is particularly instructive to observe that there are five fundamental activities that immigrant groups in the United States, and people in general, should embrace to achieve a higher standard of living. Groups that strongly embraced all five have flourished in a relatively short period of time. Groups that have not made the important connections between these fundamentals and improved living standards have languished at a lower standard of living for generations.

The following are modified versions of the previously published activities.[73]

- Be honest and obey the law.
- Work hard.
- Be self-reliant and do not accept government assistance.
- Learn to read, write, and speak excellent English; learn the local culture well.
- Obtain the most advanced education possible in a field that is in demand for employment.

The United States developed significant industrial and technological capabilities, and by the mid-1900s, workers had won better pay and working conditions. However, this development after urbanization was not necessarily a causal event, that is, increasing urbanization does not automatically increase prosperity. For example, wantonly moving thousands (or millions) of people from rural areas to urban areas can be disastrous if suitable employment opportunities are not available to match the people's skills.

Nonetheless, performing these activities has worked well in the United States and other countries since the 1800s. However, the Internet is changing this by acting as a headwind against the movement of people by enabling people in rural locations to effectively work in an urban environment without physically being in that urban environment. Further, people living in other countries with an appropriate skillset and excellent language skills are similarly enabled to work in an urban environment in a country where they do not live, effectively creating worldwide competition for employment.

By way of example, if you work directly for a company that conducts business in another language, you will be required to not only be competent, but also to read, write, and speak that language in an excellent manner. Learning and understanding the culture would be a plus. Stated differently, people at the company will be much more comfortable with you if you can communicate well and relate to them, even if you live on another continent. Conversely, communicating poorly is a distinct negative that will make your job more difficult.

Tailwinds to control global population growth include the following:

- The global population will stabilize around the year 2100.
- The birth rate will continue to fall as people living at lower standards of living understand that having fewer babies is in their best interest.
- People with lower standards of living will have the opportunity to perform the five fundamental activities that can improve their standard of living.

Perspective

For the average person, life in Malthus' world of 1798 was a life of scarcity in balance with death. Reduced child mortality, urbanization, and other factors transformed that dreadful life into an abundant 'life of life' that will be available to additional people as they adopt the fundamental activities that can improve their standard of living. However, more people living at higher standards tend to consume more resources, which increases emissions.

The birth rate should continue to fall, but the global population will increase until around 2100. The ability to compete for employment anywhere in the world while living in another city, state, or country increases competition for talent, and provides people with opportunities to improve their standard of living without moving. Increasing standards of living, regardless of location, generally increases the consumption of goods and services, which increases emissions. However, population stabilization will limit emissions if the trend of new technologies consuming the same or lesser amounts of resources continues.

The information presented in this chapter is the result of ongoing human processes that are further described in the next chapter.

Chapter 5: Rising Living Standards

May you live in interesting times. --- *Joseph Chamberlin*

Parents around the world have a lot in common, especially when it comes to eating. They coax and cajole their young children to eat as much as possible, so they grow up to be big and strong. Tactics include talking (even when the child has limited language skills), soothing, laughing, tickling, making noises, making faces, opening mouths to demonstrate, putting the food near the child's mouth, and trying to communicate how good the food tastes. I do not remember, but I suppose that my parents were no different.

What I do remember is that I eventually acquired language skills and learned that there were other places in the world that were different from where I lived. That's when the big guns came out. I was told that if I did not eat all my food, it would be packed up and sent to children who are starving in a faraway country. This tactic probably made me eat a bit more. However, it was also an introduction to the digital world.

Often, people prefer to think of people, things, events, and the like as being digital in the sense of being, for example, good or bad, short or tall, outgoing or quiet, arrogant or humble, and smart or stupid. Having only two options is often easy and, as a result, sometimes comforting. Think about that for a minute. Is it easier to describe an object as either black or white instead of a specific shade of gray? Is it easier to describe a person as good or bad instead of providing lists of pluses and minuses? Is it easier to grade a yes-or-no test question than an essay? Is it easier to describe a simple preference than the same preference with a list of exceptions and nuances?

By threatening to send my food to children in a faraway country, my parents had unknowingly divided the world into two groups. There were children who had food, and children who did not have food. Subliminally, they were suggesting that we were *rich*, and others were *poor*. This concept supports narratives such as *us and them*, the *haves and have nots*, and *the West and rest*.

---xxx---

Spoiler Alert: This chapter is not about climate change per se but, is related in the sense that it provides an understanding of some human processes underlying the transition that lifted many people out of poverty starting around 1800.

A model of the world where there are only two options is easy to understand. It is also comforting if you find yourself in the *rich, haves, the West*, or *us* groups. In the previous chapter, average people were observed to become more prosperous in the 1900s as urbanization increased, child mortality decreased, and women started having fewer babies. However, people in the countries that transitioned more rapidly enjoyed better lives sooner than people in the countries that were slow to transition. A digital model may have been valid during this stage of the transition because some countries enjoyed benefits while others did not. These groups were called *developed countries* and *developing countries*, respectively.

Income Levels

The first instinct Rosling presents is the *gap instinct*, which implies that one group is different as compared to another group when there may, in fact, be little difference between the groups. There may have been a gap between developed and developing countries at one time. However, that gap has closed significantly and overlapped, as many individuals in developing countries gradually raised their standards of living to be on par with individuals in developed countries. This reality can no longer be accurately described by only two distinct groups, but rather by multiple groups that overlap, through which people can progress as they become more prosperous. In other words, the simple digital model with only two options is not accurate and needs to be updated.

Rosling provides detailed explanations of the four income levels shown in Table 4-1. [74] The salient characteristics of these levels were found to be strikingly similar across multiple societies in different locations that had no appreciable contact with one another. Note that most people in the world live in Level 2 and Level 3 where most of their basic needs are met.

Level 1	1 billion people in 2017
Level 2	3 billion people in 2017
Level 3	2 billion people in 2017
Level 4	1 billion people in 2017

Table 5-1 Estimate of Number of People Living at Each Level in 2017 [75]

Income Level Advancement

It is possible to move up the ladder to Level 4, but it usually takes a few generations, and it is rare for one person to achieve this feat in his/her lifetime.

People at Level 1 earn less than a nominal USD 2 per day. You and your five children walk barefoot for hours each day to fetch dirty water from a well or river and gather wood on the way for the small fire used to cook the same mush served every day at every meal. You go to bed hungry if the last harvest was poor. One of your children gets sick and dies because you do not have enough money to buy medicine. This is extreme poverty, and approximately one billion people live like this. But you get lucky and sell enough of your crop to move up to Level 2.

People at Level 2 nominally earn between USD 2 and USD 8 per day. You now have some money, so you buy some plastic buckets and a bicycle to reduce the time necessary to fetch water. You wear sandals instead of going barefoot, and now you cook your food on a small stove fueled by a gas cylinder. Your children can go to school instead of fetching water and firewood. Your new electrical connection can power lights but is too unstable for a refrigerator. Mattresses are an upgrade from sleeping on the floor. Extra money can be used to purchase vegetables or maybe chickens to produce eggs. If you are lucky, you will land a factory job, earn more money, and move up to Level 3. If you are not lucky, someone in your family will get sick, and the cost of medicine will put you back at Level 1. Approximately three billion people live like this.

People at Level 3 nominally earn between USD 8 and USD 32 per day. You are working like crazy to earn more money. You now have cold running water, so no one must fetch water. You ride your motorcycle to work and cook various types of food on a two-burner stove. Your electrical connection is more reliable, so you purchased a refrigerator to store food and can serve different foods each day. The children are doing better in school because the electrical connection is better. You saved enough money so that you will not fall to Level 2 if someone gets sick. Your children graduated high school so they could get better-paying jobs. You take your family on vacation for a day at the beach. You can move up to Level 4 if you get a better job. Approximately two billion people live like this.

People at Level 4 earn more than a nominal USD 32 per day. You have hot and cold running water and cook multi-colored meals on a four-burner stove with an oven permanently connected to a gas distribution system. You purchased a car and have more than a high school education. You have enough money to be a consumer, and you try to help people in extreme poverty with a small amount of money that can make a big difference in their lives. You travel by airplane when you take vacations and eat out at least once a month. Approximately one billion people live like this.

Reading these four narratives carefully should impart an additional sense of how difficult life was for the average person prior to 1800, and how much progress has been made since. These narratives and data show that most people in the world live in the middle (Level 2 and Level 3), between the so-called rich (Level 4) and poor (Level 1). Importantly, it is often difficult for people at Level 4 to see people distinctly at the other three levels.

Figure 5-1 shows that over 60 percent of the global population was in extreme poverty prior to 1900. In the 2020s, the better part of one billion people will live in extreme poverty, so there are still many more people to be lifted out of poverty. On the brighter side, extreme poverty has been reduced from approximately 75 percent of the population to under 10 percent in a relatively short period of time.

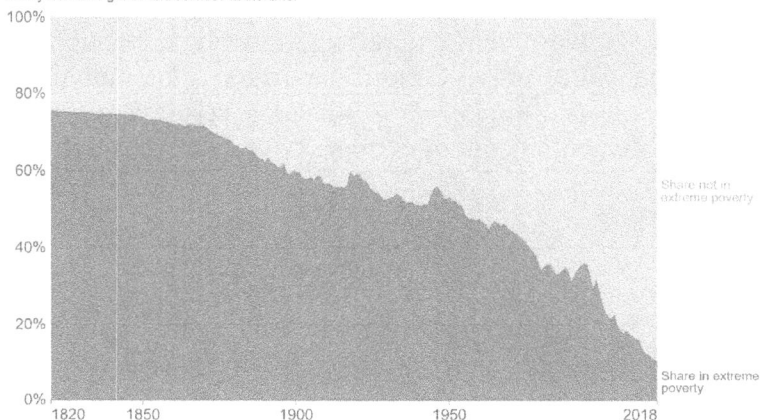

Figure 5-1 Share of population living in extreme poverty, World, 1820 to 2018 [76]

Similarly, there are approximately five billion people living at Level 2 and Level 3 who aspire to make progress and live at a higher level. Therefore, not only will the global population continue to grow until approximately 2100, but more people will increase consumption as they advance from one level to the next. Increasing populations and increasing consumption tend to increase carbon dioxide emissions. Technological advances and efforts to curb emissions present headwinds against these increases.

Government

In general, governments provide structure for society, and as such, can have an oversized influence on prosperity and daily life. Common roles of a government include creating and enforcing the rules of society, providing public services, and addressing defense, foreign affairs, and the economy. Many governments provide social services to portions of their populations.

For thousands of years, rulers either inherited their position, or ascended by force. Threats to their positions were often met with tactics that would now be considered brutal.

Governments later took on many forms, including monarchy, empire, dictatorship, socialism, Marxism, communism, republic, and democracy, some of which utilized repressive tactics to remain in power. Often, governments were and are a combination of these forms, but one or two forms usually dominate any given government.

No form of government is perfect. However, republics and democracies generally provide more opportunity for individuals to advance from Level 1 to Level 4, and as a result, have elevated much of their populations out of extreme poverty to prosperity.

Chapter 6: Energy Sources and Combustion

They paved paradise and put up a parking lot. --- Big Yellow Taxi (song by Joni Mitchell)

It is surprising how much energy can be contained in a small amount of material. We did an experiment in the chemistry lab that liberated hydrogen by reacting sodium with water in a test tube. For whatever reason, I stopped the hydrogen from leaving the test tub with my thumb. I could feel the pressure in the test tube increasing. I removed my thumb a few seconds later when a small flame started to appear. The explosion was so loud that everyone in the lab immediately stopped what they were doing and came over to see what happened. The good news was that the test tube was strong enough to withstand the explosion, and no one (especially me) was injured.

---xxx---

Various power sources have been used throughout the centuries for heat, light, and transportation. Of interest are those power sources that were in common use after approximately 1800 and their alternatives, as shown in Figure 6-1.

Global primary energy consumption by source

Primary energy is calculated based on the 'substitution method' which takes account of the inefficiencies in fossil fuel production by converting non-fossil energy into the energy inputs required if they had the same conversion losses as fossil fuels.

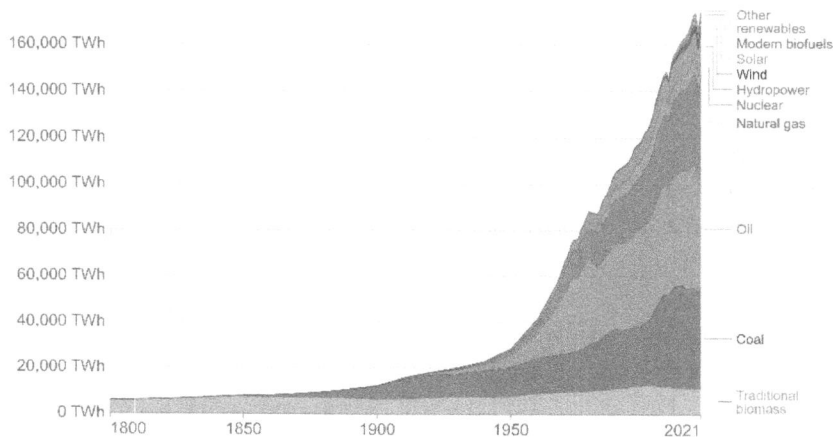

Figure 6-1 Global primary energy consumption by source [77]

Biomass, including wood, was the primary source of energy in the 1800s, when the global population was relatively small. Most people at that time lived in poverty at Level 1 and used a small amount of energy to survive, albeit inefficiently and with little or no regard for the environment. By 1900, coal provided approximately half of the world's energy. Oil, natural gas, nuclear, and hydropower were in widespread use by 2000. Wind, solar, biofuels, and other renewables were available in the 2020s.

The conditions under which these energy sources operate vary widely. Some energy sources operate at ambient temperatures and do not create heat that can warm the Earth. Other sources require operation at high temperatures, emit heat into the environment, and often emit carbon dioxide into the atmosphere.

Fossil fuels are combustible organic materials, such as oil, coal, or natural gas, derived from the remains of former life. [78] They produce both heat and carbon dioxide, which are emitted into the environment. However, **thousands of products are manufactured directly from fossil fuels.** Plant-derived molecules are generally not suitable to produce most of these products.

Energy processes are *carbon-neutral* when they emit no carbon dioxide into the atmosphere or when the net amount of carbon dioxide absorbed and subsequently released is the same. In other words, carbon-neutrality implies that there is no net gain or loss of carbon dioxide during the complete process.

Similarly, energy processes are *heat-neutral* when the amount of heat absorbed and subsequently released is the same as the amount of heat that would have been absorbed by the Earth, had it not been otherwise absorbed by the process. In other words, heat-neutrality implies that there is no net gain or loss of heat into the environment during the entire process.

These definitions are not intended to be rigorously applied because seemingly neutral energy sources can and do require ancillary processes that are not neutral. For example, generating electricity using solar cells is carbon-neutral and heat-neutral because the process does not generate heat or emit carbon dioxide, respectively. However, the ancillary processes of manufacturing, transporting, installing, and maintaining solar cells are neither carbon-neutral nor heat-neutral. Ancillary processes are ignored in this book because the amount of energy they consume is typically a small fraction of that consumed in the energy process with the caveat that this may not always be the case, such as when fuels are transported over large distances.

Energy Sources

The use of renewable energy from *biomass* (plants and animals) for cooking and heating has been a common practice for centuries. The photosynthesis process helps plants grow and store chemical energy using carbon dioxide in the air and energy from the Sun. The burning of plants releases carbon dioxide and heat that was stored as chemical energy. If the plants are replaced in kind, the same amount of heat and carbon dioxide will eventually be absorbed by the new plants, so the process is both carbon-neutral and heat-neutral over time. Cultivation of other plant species can possibly increase the speed with which both heat and carbon dioxide are absorbed.

For example, sugar cane can be harvested to produce sugar, but unused parts of the plant need to be disposed of and are often burned for their heat content. This combustion process addresses a significant waste disposal issue in a manner that is both carbon-neutral and heat-neutral, because the sugar cane plants are replaced after harvest with plants for the next crop.

However, burning down a forest where the plants that replace it absorb less carbon dioxide than the original forest is neither carbon-neutral nor heat-neutral because the net effect is the release of stored chemical energy and carbon. Replacement of forests with buildings, parking lots, and the like that cannot absorb carbon dioxide or store chemical energy creates a virtually permanent release of carbon (as carbon dioxide) and chemical energy (as heat).

Coal is a significant source of energy that is mined in many locations and processed to remove impurities. Coal is typically over 80 percent carbon plus smaller amounts of hydrogen, oxygen, nitrogen, and sulfur. Much of the activity surrounding the mining of coal occurs underground, in an environment that has been relatively dangerous for workers. However, coal fatalities in the United States have declined dramatically since 1900 with the incorporation of safety-related equipment and practices. [79] Burning coal as a fuel releases the entirety of its stored chemical energy (as heat) and carbon (as carbon dioxide) into the environment.

Crude oil is extracted from the Earth in many locations around the world but dominates economic activity in many countries in the Middle East. Crude oil is broken down (cracked) and refined to produce various fuels and petroleum products. Fuels such as gasoline, diesel, jet fuel, and heating oil release all their chemical energy and carbon (as carbon dioxide) into the environment, whereas petroleum products used to make actual products such as asphalt, tar, plastics, fertilizers, cosmetics, medicines, and the like, do not.

For calculation purposes, it is assumed that approximately 80 percent of crude oil is converted into fuels that are burned for their heat content. Extracting and burning oil from the Earth effectively removes chemical energy and carbon from deep storage. Burning oil releases the entirety of its stored chemical energy (as heat) and carbon (as carbon dioxide) into the environment, effectively adding to the amount of heat and carbon dioxide already above ground.

Hydrogen is worthy of mention as a potential fuel source that may, or may not, be viable on a large scale. Naturally occurring hydrogen was generally thought not to exist, but has been found in the United States, eastern Europe, Russia, Australia, Oman, as well as France and Mali, where a reservoir containing 98 percent hydrogen has been providing energy to the city of Bourakébougou for more than a decade. Research and development activities focus on locating hydrogen reservoirs, determining their size, and developing processes to purify their contents in an economically viable manner [80] in addition to investigating technologies to produce hydrogen economically with low carbon dioxide emissions. [81]

Methane is by far the largest single component in *natural gas*. However, the composition of natural gas varies with location. Extracted from wells located in various locations around the world, natural gas is usually burned for its heat content, thereby releasing its chemical energy and carbon (as carbon dioxide) into the environment.

Heat from *nuclear* reactions has been used commercially as an energy source to generate electricity since the 1950s. There were over 400 nuclear power plants worldwide in the 2020s that produced a significant amount of electricity. Nuclear energy plants do not emit carbon dioxide but do emit heat into the environment.

Hydropower plants produce electricity from the natural flow of water, which is used to spin a turbine. These plants often contain a dam that harnesses power upstream, from which the water flow into the turbine can be controlled. Early factories were often located near rivers where mechanical hydropower could be harnessed to directly operate machines. Hydropower plants do not emit carbon dioxide or release heat into the environment.

Wind power has been used for centuries, such as to pump water and power ships. In the early 2000s, wind turbines gained popularity for producing electricity. Wind turbines do not emit carbon dioxide or release heat into the environment, but they are known to kill birds that fly into their blades. Wind energy is intermittent and is not available when the wind is not blowing.

Solar panels that directly convert sunlight into electricity were in wide use in the 2020s but represented only a small amount of electricity production. Rooftop solar systems have been installed on single-family houses, while large-scale systems have been installed in solar farms. Locations closer to the equator generally produce more electricity than those farther away. Solar energy is intermittent, producing less electricity on cloudy days and no electricity when the sun is not shining. Solar panels do not emit carbon dioxide or release heat into the environment.

Biofuels are fuels that are derived from plants and animals, and often emulate fuels derived from oil or natural gas. Some biofuels have the disadvantage of displacing food production, which tends to increase food prices and sometimes reduce the food supply. Examples include ethanol produced from sugar cane or corn, and bio-jet fuel made from landfill waste. Biogas derived from animal waste has an added advantage of converting waste into useful fuel. Some biofuel processes have the additional benefit of reducing the amount of waste that would otherwise need to be disposed of or destroyed.

Geothermal plants utilize heat from reservoirs of hot water or steam below the surface to generate electricity, heat, and cool. Their application is limited to locations where thermal energy can be readily extracted from the Earth. Geothermal energy is carbon-neutral and can operate reliably regardless of weather conditions.

There are many other alternative sources of energy that are practiced on a small scale or are not yet practical, such as tidal and wave energy. The technologies associated with these energy sources are often carbon-neutral and heat-neutral.

Energy sources change over time, so there are likely sources of energy that are currently unknown and undreamt of but will one day be discovered, developed, and become viable. Simply put, we do not know what we do not know.

Combustion

Many energy sources require the burning of a fuel-air mixture that produces carbon dioxide, water vapor, and heat, where the heat is used in the process, as illustrated in the following series of chemical reactions. The line below each reaction shows the mass needed to burn one mole (6.022×10^{23} molecules) of fuel. Nitrogen in the air is not shown in the equations, because it is not part of the reaction. British thermal units (BTU) and kilojoules (kJ) are measures of heat content. Molecular weights are approximate and complete combustion is assumed, so 100 percent of the fuel is combusted. Discussions and calculations assume that the fuel is completely combusted.

The following chemical reaction describes the combustion reaction of hydrogen and oxygen in the air. Burning 4 grams of hydrogen produces 36 grams of water vapor but does not produce any carbon dioxide. Other fuels, such as ammonia, contain no carbon and similarly produce no carbon dioxide. However, hydrogen and ammonia are largely manufactured gases, and not energy sources. However, the exploitation of natural hydrogen reservoirs could potentially make hydrogen a significant source of energy.

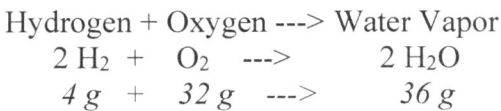

$$\text{Hydrogen} + \text{Oxygen} \dashrightarrow \text{Water Vapor}$$
$$2\,H_2 + O_2 \dashrightarrow 2\,H_2O$$
$$4\,g + 32\,g \dashrightarrow 36\,g$$

The following equation describes the chemical reaction commonly used to describe the combustion reaction for natural gas, assuming that the natural gas is pure methane. The mass of the carbon dioxide produced is 2.75 times (44/16) the mass of the methane combusted.

Methane + Oxygen ---> Carbon Dioxide + Water Vapor
$$CH_4 + 2\,O_2 \quad ---> \quad CO_2 \quad + \quad 2\,H_2O$$
$$16\,g + 64\,g \quad ---> \quad 44\,g \quad + \quad 36\,g$$

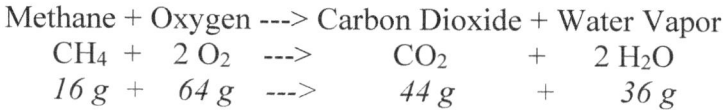

The following equation describes the chemical reaction that can be used to describe the combustion reaction of coal, assuming that the coal is pure carbon. The mass of the carbon dioxide produced is 3.67 times (44/12) the mass of the coal combusted.

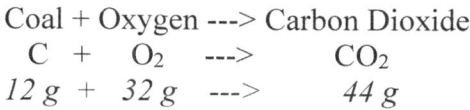

Coal + Oxygen ---> Carbon Dioxide
$$C + O_2 \quad ---> \quad CO_2$$
$$12\,g + 32\,g \quad ---> \quad 44\,g$$

Of interest is the amount of carbon dioxide produced per unit heat release, such as per million BTU, as shown in Figure 6-2 for various fuels.

Pounds of CO_2 emitted per million British thermal units (Btu) of energy for various fuels:

Coal (anthracite)	228.6
Coal (bituminous)	205.7
Coal (lignite)	215.4
Coal (subbituminous)	214.3
Diesel fuel and heating oil	161.3
Gasoline (without ethanol)	157.2
Propane	139.0
Natural gas	117.0

Figure 6-2 Carbon Dioxide Emitted per Million BTU of Various Fuels [82]

The following is a summary of select attributes of the various energy sources, noting that ancillary processes consume other energy sources that can produce carbon dioxide and heat.

- Energy sources that produce no carbon dioxide and no heat
  - Hydropower, solar, and wind
- Energy sources that produce no *net* carbon dioxide and no net heat
  - Cultivated biomass energy sources where new plants replace harvested plants.
- Energy sources that produce heat but no carbon dioxide
  - Hydrogen (naturally occurring)
  - Nuclear
- Energy sources that produce carbon dioxide and heat
  - Natural gas
  - Gasoline and diesel
  - Coal

A cursory inspection of the above items reveals that the energy sources in the first three solidly bulleted items emit no carbon dioxide. The fossil fuel energy sources listed in the fourth bulleted item should be considered in the order presented to reduce the amount of carbon dioxide produced. On a heat release basis, gasoline and diesel fuels emit approximately 35 to 40 percent more carbon dioxide than natural gas, while coal emits approximately 100 percent more.

Adding granularity, examination in conjunction with Figure 6-1 reveals that the traditional biomass, coal, oil, natural gas, nuclear, and hydropower energy sources represent approximately 95 percent of energy consumption around 2020. These energy sources showed decreased consumption around 2020 during the COVID-19 epidemic but did not recover to previous levels in the early 2020s, perhaps because subsequent increases in energy consumption were offset by wind, solar, modern biofuels, and other energy sources that are in the first and second solid bulleted items. This shift to carbon-neutral and heat-neutral energy sources in the early 2020s could potentially be the start of a new trend.

Carbon and Thermal Footprints

A carbon footprint is the total greenhouse gas (GHG) emissions, expressed as carbon dioxide equivalent (CO_2e) caused by an individual, event, organization, service, place, or product. [83] Processes that consume fossil fuels produce carbon dioxide and have a carbon dioxide footprint. For example, a person's carbon footprint includes the amount of greenhouse gases emitted to breathe, build a house, travel by car, motorcycle, bus, train, or air, and produce the goods and services that the person uses or consumes. Footprint calculators are available on the Internet that can provide an appreciation of the carbon footprints associated with many everyday goods and services.

While a detailed calculation of a carbon footprint might prove interesting, its applicability to emissions is minimal due to the large differences in the scale necessary to satisfy demand. For example, turning off the lights in unoccupied rooms and reducing thermostat settings at night are actions that might reduce a person's carbon footprint by 5 or 10 percent, and maybe more. This accomplishment might be a big deal for an individual however it is nothing more than a very small drop in a very big bucket when dealing with a global issue. Stated differently, there are much bigger fish to fry that can potentially affect emissions much more significantly.

A thermal footprint is defined herein as waste heat that is rejected to the environment. Processes that generate heat, such as the burning of fossil fuels or generating electricity from nuclear energy have a thermal footprint. For example, hot gases exiting a home furnace and water heater are routed up a chimney and warm the atmosphere. Similarly, hot gases produced by an oven are dispersed into the air and affect the environment by warming the kitchen. Both processes emit heat, and therefore both have thermal footprints.

The next chapter discusses the utilization of fossil fuel and nuclear energy sources in processes that have a carbon footprint, thermal footprint, or both.

Chapter 7: Steam Boilers and Electricity

Some consultants are like the bottom half of a double boiler: They get all heated up but don't know what's cooking. --- Origin unknown

One sunny morning, the boiler house supervisor and I were in the control room discussing an operational matter when a forklift driver came in and excitedly told us that there was a fire in the incinerator. A glance at the temperature controller showed that incinerator operation was stable at its usual 982 degrees Celsius (1800 degrees Fahrenheit). The supervisor and I exchanged smirks, after which the supervisor assured the driver that there was a fire in the incinerator.

The forklift driver would have none of that and insisted that there was a fire in the incinerator. Realizing that the driver was not going to let us disregard him, we made our way out the front door of the boiler house and around another boiler to inspect the incinerator. When we came around the corner, we saw an open flame above the incinerator shell that was approximately three meters high and looked like it could have been part of a special effect in a movie.

The supervisor radioed the control room and ordered the operators to immediately pull the fume feed to the incinerator, and the flame went out. Inspection revealed that a crack in the refractory allowed hot gas to weaken the incinerator shell, causing a nozzle to shift. This shift caused the nozzle's gasket to fail and leak combustible gases, which were ignited by the hot metal shell.

The size of the flame was impressive given that the visible fire was burning less than 5 percent of the total energy being supplied to the incinerator at the time. It gave us an appreciation for how much energy can be contained in a relatively small amount of fuel.

---xxx---

Multiple forms of energy are consumed in numerous ways that are designed to meet the specific needs of various industries and people. The specific needs of each process determine the details of each installation.

The industrial sector accounts for a large portion of global energy use. Some of this energy is consumed directly in processes, such as using coke to make steel. However, many industries use boilers to generate steam to heat their processes. Utilities use steam from boilers to generate electricity that is used to operate industrial equipment and for public use. The sheer magnitude of energy consumption in this sector drives constant innovation to improve energy efficiency, reduce energy consumption, and reduce costs to stay competitive.

Interestingly, industry goals of reducing energy consumption and saving money, and environmental goals of reducing the release of heat and carbon dioxide emissions are in strong alignment with one another. In this sense, industry's success is also an environmental success, and vice versa.

Steam Boilers

Many manufacturing plants have boilers that burn fossil fuels to generate steam, which is used to provide heat for processes. Fossil fuels such as natural gas, oil, or coal, are typically mixed with a desired amount of combustion air and burned to produce hot products of combustion that largely contain carbon dioxide, water vapor, nitrogen present in the combustion air, some oxygen that was not consumed in the combustion process, and relatively small quantities of other gases. Heat in the products of combustion is transferred to water that boils to produce steam, which is transported in pipes to equipment where it is utilized in processes. The cool products of combustion are dispersed into the atmosphere.

Some of the heat released is used to form water vapor, heat the nitrogen in the combustion air, escape as heat losses, and operate ancillary systems, such as an atomizing steam system. Therefore, some of the heat released from the fuel is lost to the environment and cannot be effectively used to generate steam. Boiler efficiencies can range from approximately 70 to 95 percent depending on design, fuel, and operating conditions, among other factors. Boilers produce carbon dioxide and release heat into the environment.

This general boiler description does not address the many equipment variations and operating conditions that are encountered in actual boiler installations, including boilers that burn multiple fuels, specific fuels, solids, and waste materials.

Equipment that utilizes geothermal energy from the Earth can also be designed to generate steam.

Reliable Electricity

Reliable electricity is electricity that is produced from energy sources and processes that are always available, such as coal, oil, natural gas, nuclear, geothermal, and hydropower. Electricity sources that produce electricity intermittently, such as solar and wind, are not considered reliable. This is because solar sources do not produce electricity at night or on a cloudy day, and wind sources do not produce electricity when the wind is not blowing.

It is extremely important that reliable electricity sources dominate electrical grids. Failure to be reliable makes the grid prone to outages that typically cause more than just an inconvenience. For example, electrical outages can cause the lights to go out, food in people's refrigerators to spoil, elevators with people to stop mid-floor, building heat in the winter to fail to operate, businesses to be unable to process sales payments, and threaten the lives of people dependent on life-sustaining machines.

Electricity From Fossil Fuel and Nuclear Energy Sources

Electrical energy is a convenient and versatile energy source because, once connected to the grid, it is relatively easy to transport to other locations and electric consumers. Electricity can be generated from various energy sources using various processes.

Fossil fuel boilers in power plants typically burn natural gas, oil, or coal to produce high-pressure, high-temperature steam that is used to operate a steam turbine that rotates a generator that produces electricity. Nuclear power plants operate in a similar manner, except that heat from a nuclear reaction is used to produce high-pressure, high-temperature steam.

The efficiency of a power plant can be calculated from its heat rate. In 2021, the average operating heat rates for various fuels in British thermal units per kilowatt hour (BTU/kWh) were coal (10,583), natural gas (7687), oil (11,223), and nuclear (10,429). [84] One kilowatt is equivalent to 3413 BTU/h, so the efficiency with which a fuel produces electricity can be calculated as 3413 divided by the heat rate. The approximate efficiencies are coal (32.2 percent), natural gas (44.4 percent), oil (30.4 percent), and nuclear (32.7 percent). In addition, approximately five percent of the electrical energy transmitted is lost due to transmission and distribution losses [85] so more accurate estimates of the overall electrical efficiency from source to user are coal (30.6 percent), natural gas (42.2 percent), oil (28.9 percent), and nuclear (31.1 percent).

One immediate observation is that power plants powered by natural gas are far more efficient than those that utilize coal, oil, or nuclear fuels. This is primarily due to their design, so converting oil-burning and coal-burning boilers to burn natural gas would reduce emissions, but not necessarily change their overall efficiency without additional design changes.

Power plants that burn natural gas typically incorporate a combined-cycle process to enhance electricity generation. Higher efficiency is achieved by first using a gas turbine to produce electricity, and then routing its hot exhaust gases to a heat recovery steam generator (HRSG) that produces steam to operate a steam turbine that generates additional electricity. If so designed, some of the steam can be used to enhance electricity production in the gas turbine.

Another observation is that producing electricity is a relatively inefficient process, because a significant amount of heat is wasted and rejected into the environment. This is largely because steam is condensed after extracting as much energy as possible to produce electricity. For example, if each pound of steam leaving a boiler contains 1500 BTU, and 950 BTU are lost due to condensation after spinning a turbine, approximately 63 percent of the energy in the steam is rejected to the environment by this process.

Nonetheless, information about some electrical appliances, such as electric heaters, may claim efficiencies approaching 100 percent. This would be correct if efficiency were measured from the heater plug to the output of the heater. However, energy efficiency should be measured from energy sources such as coal, oil, natural gas, or nuclear, to the output of the heater. As such, the significantly lower efficiencies calculated above should be considered.

Cogeneration

Cogeneration is the sequential use of energy that enables waste energy from one part of a process to be used productively in another part of the process. The goal is to make more efficient use of the fuel that is consumed.

For example, heat is removed from a vehicle's motor to keep it cool. Some of this waste heat is used to heat the interior of the vehicle, while the remainder is rejected to the environment. This process consumes the same amount of gasoline to operate both the engine and the heater as it would if it operated the engine only. Another example is using hot gas leaving a piece of equipment to preheat cooler gas entering the equipment. Both processes reduce emissions because cogeneration enables them to consume energy sequentially.

In practice, a fossil-fuel-fired turbine can be used to generate electricity and produce hot gases with energy that can be used to generate steam for consumption in a nearby process instead of being wasted as would occur in a traditional utility boiler. **Cogeneration is significantly more efficient and less expensive than purchasing electricity and producing steam onsite.**

Electricity From Renewable Energy Sources

Geothermal, hydropower, solar, tidal, wave, and wind power plants generate electricity directly using energy harvested from nature. Geothermal energy derived from hot water or steam below the surface of the Earth can be used to generate electricity and for heating. Hydroelectric plants that use the energy in flowing water to rotate turbines that generate electricity are reliable and have been in use since the late 1800s. Solar electric farms, which use arrays of solar cells to directly convert the energy from the Sun into electricity, have been in use since the late 1900s. Tidal electric plants harness the movement of water during the tides, while wave electric plants harvest energy from waves. Wind farms generally contain multiple wind turbines that turn electric generators to make electricity.

Each renewable energy source has its limitations. Geothermal plants are generally limited to locations along the major tectonic plate boundaries, where most volcanoes are located. Hydropower plants sometimes reduce or halt electrical production for extended periods of time under drought conditions, or when the amount of water available upstream, such as in a reservoir, is diminished. Solar and wind energy are not predictable because they do not produce electricity when the Sun is not shining and the wind is not blowing, respectively. Wave energy can vary with surface conditions. Tidal energy is more predictable than solar, wave, and wind. However, it is not widely in use.

Of these renewable energy sources, only geothermal and hydroelectric energy are reliable and have a long history of operation. The other renewable technologies would benefit from electrical energy storage systems.

Electricity From Biofuel Energy Sources

Biofuels are fuels that are produced from biological sources such as animals, plants, and algae. Examples of liquid biofuels include aviation biofuel, biodiesel, ethanol, olive oil, and whale oil, some of which are made from other energy sources such as sugar cane, corn, and other plants. Solid biofuels include agricultural crops, municipal solid waste, wood in various forms, and the waste generated from these sources.

Biogas is a renewable fuel gas that is produced using microorganisms to break down organic matter in the absence of oxygen. Suitable organic raw materials include food waste, animal waste, agricultural waste, municipal waste, sewage, and wastewater. The composition of biogas varies, but it typically contains over 50 percent methane, so it can be burned for its heating value or used to generate electricity.

The heat and carbon dioxide released by burning biofuels are derived from products of nature and not liberated from carbon that is more permanently stored in the Earth, such as coal, oil, and natural gas. In this sense, biofuels are renewable. However, burning biofuels is subject to the same inefficiencies as burning fossil fuels. However, the presence of water and inert materials in biofuels can significantly reduce combustion efficiency.

Renewable biofuels offer carbon-neutrality and heat-neutrality, plus a pathway to effect energy cost savings while destroying waste that could otherwise require costly disposal. Research is ongoing to develop viable processes to convert biomass into biofuels.

Electricity: Carbon Footprint

Electricity generated from renewable and biofuel energy sources does not consume fossil fuels, so they have no net carbon footprint. However, ancillary processes can have carbon footprints, depending on the process and its energy source.

From Figure 6-2, the amount of carbon dioxide emitted per million BTU is approximately coal (215 pounds), natural gas (117 pounds), oil (159 pounds), and nuclear (0 pounds). The efficiency with which electricity is generated and delivered to the customer was previously shown to be approximately coal (30.6 percent), natural gas (42.2 percent), oil (28.9 percent), and nuclear (31.1 percent).

Therefore, the amount of carbon dioxide emitted to generate 1 megawatt-hour (MWh) of electricity is approximately coal (2398 pounds), natural gas (946 pounds), oil (1878 pounds), and nuclear (0 pounds). For example, coal releases approximately 2398 pounds of carbon dioxide per MWh of electricity delivered to the customer. [B] Calculations for the other fuels are similar.

By observation, providing the same amount of electricity to a customer by **burning oil releases more than two times the amount of carbon dioxide as compared to burning natural gas. Approximately 2.54 times as much carbon dioxide is released when burning coal instead of natural gas.**

In general, **burning fossil fuels emits large amounts of carbon dioxide**.

Electricity: Thermal Footprint

Electricity generated from renewable energy and biofuel sources is considered to have no net thermal footprint. However, ancillary processes can have a thermal footprint, depending on the process and its energy source.

Repeating from above, in 2021, the average operating heat rates for various fuels in British thermal units per kilowatt hour (BTU/kWh) were coal (10,583), natural gas (7687), oil (11,223), and nuclear (10.429). [86]

The amount of energy that is rejected or lost to the atmosphere is the heat rate less the one kilowatt-hour (3413 BTU) of electricity that is delivered to the customer. Therefore, the thermal footprints in BTU per kilowatt were coal (7170), natural gas (4274), oil (7810), and nuclear (7016). For example, coal rejects (10,583 - 3413), or approximately 7170 BTU per kWh of electricity delivered to the customer. Calculations for the other fuels are similar.

[B] (3.413 BTU/hour / Watt) * (215 pounds per million BTU) / (0.306 efficiency)

Stated succinctly, **burning fuel to generate electricity is a relatively inefficient process.**

Chapter 8: Energy Consumers

*First get your facts straight. Then distort them at your leisure. ---
Mark Twain*

Some years ago, I was staying at a hotel and planning to walk to a local restaurant to eat dinner. It looked like it would rain, so I asked the receptionist for an umbrella. The reply was, "From me?" I was thinking about saying that I wanted it from another person over there, which is why I asked you over here, but I answered yes. Then I was asked if I wanted it now. I was thinking about saying that I wanted it next Tuesday at 3:30 in the morning, which is why I asked for it now, but I answered yes.

As petty as this might seem, simple interactions can quickly become complicated.

---xxx---

Everyone is entitled to make the occasional mistake, oversight, or even blunder. However, consequences often occur when they happen repeatedly. Less serious, but still significant, are errors of omission. Have you ever received a response that addressed only three or four of the five numbered questions in your email? It is annoying to write back and ask for the answers to the other questions, but not critical.

The interrelationships between energy sources and energy users are relatively straightforward in theory but complicated in practice. Figure 8-1 is a spaghetti chart that shows the amounts of each energy source on the left and the energy consumption in four sectors of the economy on the right. The lines from left to right depict the amount of energy leaving each source and going to the various sectors, and the percentage of consumption by each sector on the right.

For example, the uppermost line shows that 59 percent of the petroleum energy produced is consumed by the transportation sector, which represents 90 percent of the energy consumed by the transportation sector. Relationships with the electric power sector are somewhat different because this sector consumes energy from various sources and supplies electricity to various sectors.

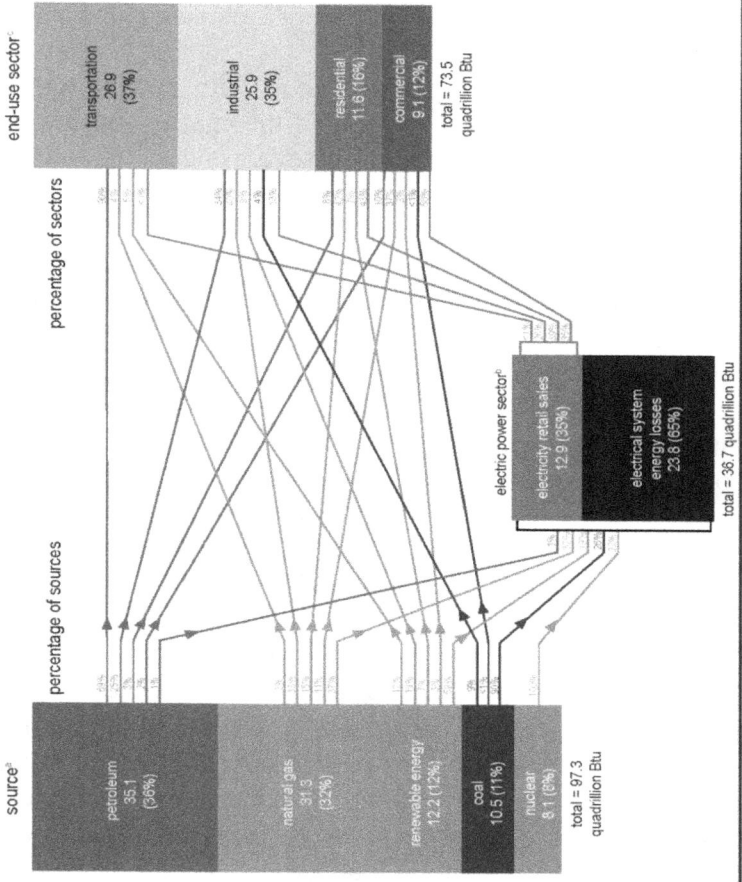

Figure 8-1 U.S. Energy Consumption by Source and Sector, 2021 [87]
(If not legible, visit https://www.eia.gov/totalenergy/data/monthly/pdf/flow/total-energy-spaghettichart-2021.pdf)

It might be helpful to bookmark, copy, or print Figure 8-1 because it will be referenced and analyzed in subsequent chapters.

Electricity (and the Marginal Megawatt)

The electric power sector uses various energy sources to generate electricity for customers in other sectors of the economy that is transported via transmission and distribution lines. Per Figure 8-1, the approximate percentages of energy sources consumed to generate electricity in the United States include coal (26 percent), natural gas (32 percent), nuclear (22 percent), and renewables (19 percent).

The overall efficiency associated with generating electricity is relatively low, so well over half of the energy consumed is lost and rejected to the environment. However, one exception is combined-cycle natural gas-fired power plants, which operate significantly more efficiently.

Electric power plants are typically large, centralized facilities that have been designed to operate for many decades. As such, it can take decades for new technology to be significantly integrated into operating power plants.

Generating plants and consumers in North America are interconnected via the electricity grid system. As electricity consumer demand varies throughout the day, an electricity bidding system determines which electrical generating plants will operate, and at what load levels. Additionally, these so-called *marginal megawatts* can be supplied from almost any location on the grid due to its high level of integration.

In practice, nuclear and hydroelectric power plants typically operate near their maximum outputs, and wind and solar energy sources produce whatever amount of electricity they can at that time. Therefore, marginal megawatts are usually generated in fossil-fuel electric plants that can be ramped up and down as needed to provide reliable power. Figure 8-1 shows that almost no electricity is produced using oil, so marginal megawatts will usually be generated using a combination of natural gas and coal.

This means that the marginal megawatt, that is, the megawatt that is needed when an additional load turns on, is causing a gas-fired power plant or coal-fired power plant to increase its output to produce that extra megawatt for the grid.

As previously discussed, power plants that burn coal have a relatively low thermal efficiency, a relatively high thermal footprint, and the highest carbon footprint of all fossil fuels. If the electricity to charge an electric vehicle is produced by a coal plant, the electric vehicle might produce more emissions than the gasoline-burning vehicle it replaces. This is also possible when natural gas is used to generate electricity.

Transportation

The transportation sector includes vehicles whose primary purpose is to transport people and/or goods from one physical location to another. Transportation vehicles derive over 90 percent of their energy from petroleum products and consume approximately 69 percent of their fuel from petroleum sources.

Transportation vehicles require a portable energy source with a high energy density. Stated differently, vehicles need to store a large amount of energy in a small volume, which is one of the reasons why gasoline and diesel fuel are so prevalent.

Vehicles are routinely replaced every few years, so design and technology changes can be substantially implemented within a few years after their introduction.

Industrial

The industrial sector contains facilities and equipment used for producing, processing, or assembling goods, including manufacturing, agriculture, mining, and construction. Energy is used to heat and cool processes and operate machinery. However, fossil fuels are sometimes used as direct raw materials for the process. Approximately 25 percent of oil energy sources, 33 percent of natural gas, 19 percent of renewable energy, 9 percent of coal, and 26 percent of the electricity generated are consumed in industrial facilities in the United States.

Industrial plants are typically custom designed using proprietary technology but are flexible enough to accommodate some changes and expansions over time. Upper management almost always requires technical and economic justification to secure funds for other than small process changes. Justification for the integration of new technology into existing processes can take a lot of time and effort. Justifying the replacement of existing processing units is typically more difficult unless it is inefficient as compared to its replacement, or difficult to maintain.

Residential

The residential sector consists of living quarters for households, space heating, water heating, air conditioning, lighting, refrigeration, cooking, and operating appliances. Approximately 7 percent of renewable energy, 15 percent of natural gas, and 39 percent of the electricity generated are consumed in residential settings in the United States.

Most household items are upgraded or replaced every so often. For example, a blender might last 3 to 5 years, so it would not take long to replace it. However, a heating, ventilating, and air conditioning (HVAC) system can last 20 years or more, so homeowners tend to delay upgrading to a new, expensive system until it is necessary, even though the new system may be significantly more efficient than the one it replaces.

Commercial

Commercial facilities consist of businesses, government buildings, and other private and public buildings and institutions. Common uses of energy include space heating, water heating, air conditioning, lighting, refrigeration, cooking, and operating other equipment in nonresidential buildings. Approximately 11 percent of natural gas, and 35 percent of the electricity generated are consumed in commercial facilities in the United States.

The functions required in commercial facilities are similar to those required in residential buildings. However, equipment in commercial buildings is typically larger and somewhat more complex.

Chapter 9: Greenhouse Gases

It just goes to show you, it's always something - if it ain't one thing,
it's another.
Roseanne Rosannadanna (played by Gilda Radner)

It must have been in the fourth or fifth grade when my public-school class made a field trip to the Museum of Science and Industry (MSI) in Chicago, Illinois. I returned so excited that I convinced my parents to take me again, and years later, I visited again with my wife. But I digress.

The most memorable exhibits were a periodic table mounted on a huge wall showing all the chemical elements, a tour of an actual German submarine captured intact during World War II, and a visit to a coal mine.

Since 1933, Coal Mine has taken MSI guests down the mineshaft, along the rails, and through the walls of a true-to-life coal mine.

It was the Museum's very first exhibit, and it's been a guest favorite ever since. You'll dig this descent into a "working" coal mine, an engaging tour of mining methods and machinery through the years. Step on the hoist and ride down into an experience with an atmosphere so real, you may start wondering where to punch your timecard.

Touring Old Ben No. 17—a mine relocated from southern Illinois—immerses you in the vast and sometimes perilous work of mining for coal. The scene is set as the facilitator narrates your cage elevator ride into the mine. See in a flash how Davy lamps reveal unseen methane dangers. Board the work train to experience a coal miner's "daily commute." See working examples of how extraction machinery has evolved from the pickaxe to longwall machines that can carve out football field-sized sections of shale.

A subterranean tour for the senses, with the real-life rumble, sounds and feel of working in a coal mine. Nothing brings home the work of mining like seeing and hearing the real, working equipment once used in active mines. Gain a new perspective on energy as you witness the mechanical might and engineering ingenuity that goes into harvesting coal. You may just see light switches in a whole new way. As much as coal mining has changed since 1933, the classic experience of a Coal Mine visit remains one-of-a-kind. [88]

We donned hardhats during the tour and saw various pieces of mining equipment. But years later, I remember the Davy lamp, which warned miners when there was bad air in the mine. It had an enclosed flame that changed in size and intensity when flammable gases were present and when the oxygen content in the mine was low, indicating that it was time for the men to leave for a safer location with better air. The Davy lamp effectively replaced the practice of bringing canaries into the mine that would chirp if all was well and stop chirping if the air in the mine went bad.

---xxx---

Carbon dioxide may possibly be the proverbial *canary in the coal mine* because combustion, which is the burning of fossil fuels, releases carbon dioxide into the atmosphere where it can accumulate. Carbon dioxide has been in the atmosphere for billions of years, playing an integral role in the photosynthesis process that enables plants to grow and store chemical energy using energy from the Sun. Stated differently, carbon dioxide may be like food for plants, but too much can harm humans and other forms of life.

Carbon Dioxide Concentration

Figure 9-1 shows that between 50 and 400 million years ago, the concentration of carbon dioxide in the atmosphere varied between approximately 200 and 2000 ppm. Variations between 1 and 50 million years ago were smaller, during which the concentration reached over 1000 ppm. In contrast, the concentration of carbon dioxide has been relatively steady from approximately 1 million years ago until recently.

In general, values on the horizontal axis and vertical axis of graphs are *linear*, so the same horizontal and vertical distance correspond to the same time and value, respectively. However, the concentration of carbon dioxide in the atmosphere over wide time periods and varying degrees of detail can be shown on one graph if the axes are *logarithmic*, such as in Figure 9-1. In particular, the time scales on the horizontal axis show time in hundreds of millions, millions, thousands, and hundreds of years, with more granularity at the right. Carbon dioxide is shown from 100 ppm to over 2000 ppm, where the spaces between 100 and 200, 500 and 1000, and 1000 and 2000 are the same because each represents a doubling of the value. Presenting data with linear and logarithmic scales can be appropriate, depending on the nature of the data. However, the person utilizing the graph must be cognizant of its scale to properly interpret the data.

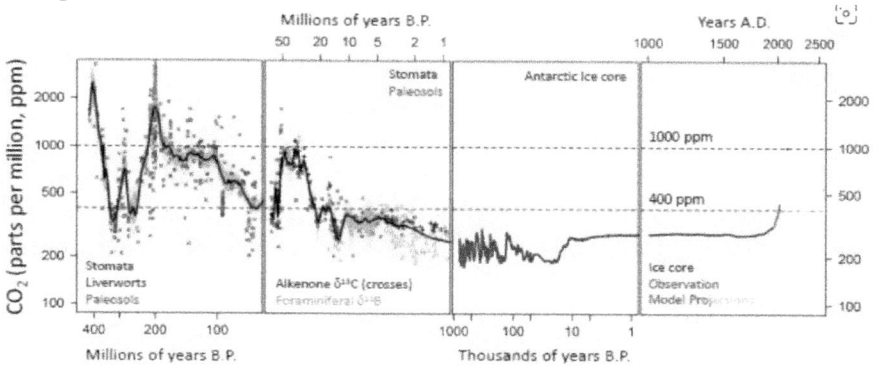

Figure 9-1 Carbon Dioxide Levels in the Atmosphere (400 million years) Without Projections [89]

What is not shown is the context of the data. The dinosaurs came and went, as did many other species of plants and animals. The Earth evolved over these 400 million years due to natural processes such as wildfires, animal respiration and decomposition, volcanic eruptions, and the weathering of carbonate rocks. In other words, the Earth was not the same tens and hundreds of millions of years ago as it is now. Therefore, the high carbon dioxide concentrations on the left side of the graph may not be relevant when analyzing carbon dioxide concentrations at the present time.

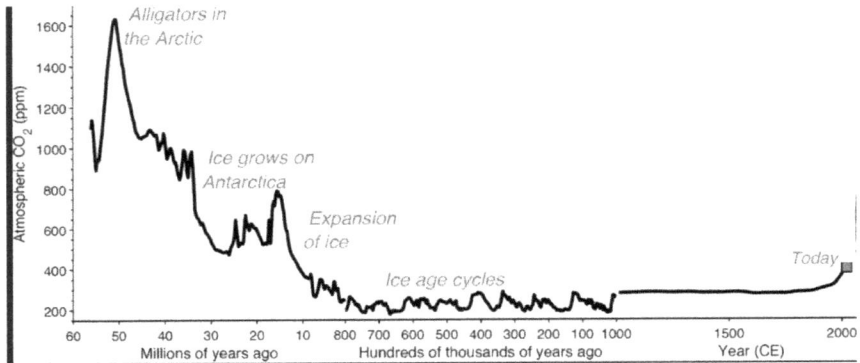

Figure 9-2 Carbon Dioxide Levels in the Atmosphere (60 million years)
Without Projections [90]

Figure 9-2 contains more granularity and shows that carbon
dioxide levels have been elevated in the past. The annotations
indicate that the Earth was very different when alligators were in the
Arctic and when ice was being formed in Antarctica. Nonetheless,
the data shows that the concentration of carbon dioxide in the
atmosphere was stable for approximately one million years and quite
steady for approximately 1000 years. Note that the horizontal axis is
logarithmic, and the vertical axis is linear but starts at 200 ppm.

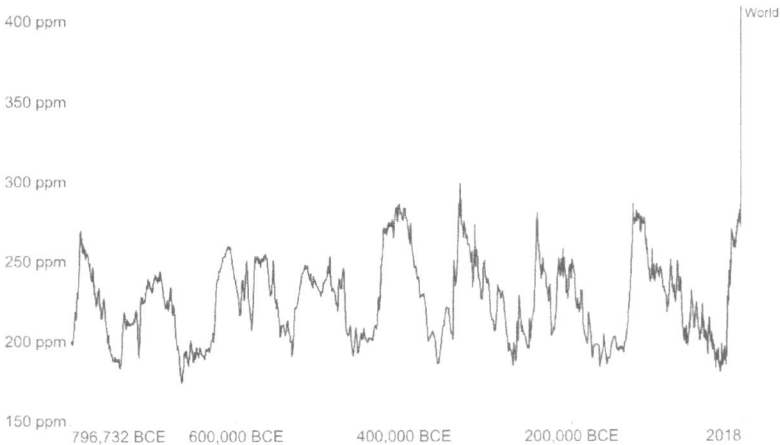

Figure 9-3 Carbon Dioxide Levels in the Atmosphere (80,000 years) [91]

Figure 9-3 has linear scales that are more granular and start at 150 ppm, making fluctuations appear more prominent, even though the carbon dioxide concentration in the atmosphere varied between approximately 180 and 300 ppm for 800,000 years. This pattern is typical of a process that is in relative balance. Stated differently, the amount of carbon dioxide generated and utilized varied somewhat over time but was approximately the same for 800,000 years. There was no human influence during the first 600,000 years, and minimal human influence until recently.

10,000 Years of Carbon Dioxide

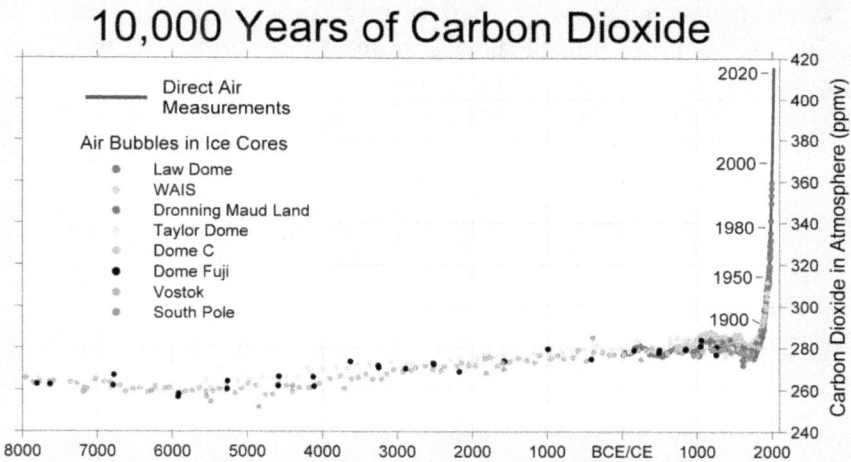

Figure 9-4 Carbon Dioxide Levels in the Atmosphere (10,000 years) [92]

Figure 9-4 is even more granular and shows that the concentration of carbon dioxide in the atmosphere has been approximately 280 ppm for the last 3000 years, until starting to rise around 1800 and accelerating after around 1910. Note that the vertical axis of 240 ppm to 420 ppm magnifies the changes and gives the impression that they are large because data points appear from the bottom to the top of the graph. That said, experience as a process control engineer would lead one to suspect that an event or series of events caused the rise in the early 1900s.

The increase in concentration starting around 1800 coincides with when humans started burning significant amounts of fossil fuels. Plants may have absorbed some excess carbon dioxide at first, but it can be theorized that as man burned more and more fossil fuels, the plants could not keep up with the carbon dioxide produced, and the amount of carbon dioxide in the atmosphere increased. While this theory makes sense and is generally accepted, the increase in the amount of carbon dioxide in the atmosphere could be coincidental and not causal. In other words, the possibility of coincidence should be approached with an open mind.

The rise since 1800 may not appear to be that much different when compared to previous changes in concentration. However, note that the increase occurred in only 200 years, while the previous changes occurred over hundreds of thousands of years. In general, this relatively fast response often occurs in processes after a process-altering event occurs or when a new process starts. Examples include an asteroid hitting the Earth and the burning of fossil fuels, respectively. The latter has been suspected of causing carbon dioxide levels in the atmosphere to rise.

Global atmospheric CO2 concentration

Atmospheric carbon dioxide (CO₂) concentration is measured in parts per million (ppm). Long-term trends in CO₂ concentrations can be measured at high-resolution using preserved air samples from ice cores.

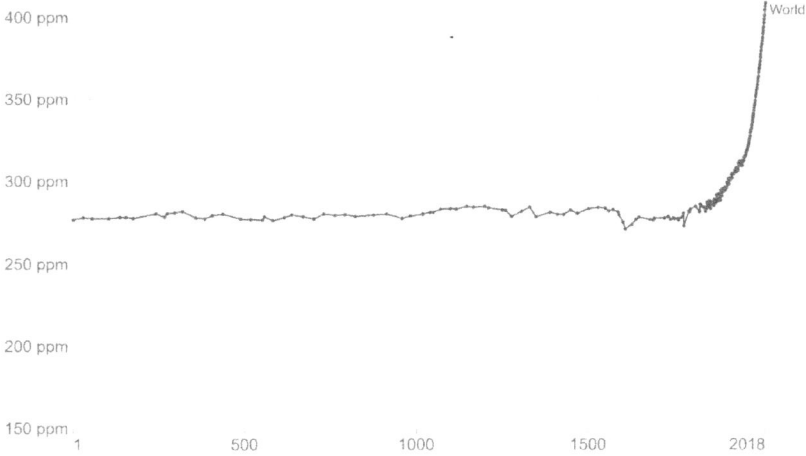

Figure 9-5 Carbon Dioxide Levels in the Atmosphere (2000 years) [93]

Figure 9-5 shows that the concentration of carbon dioxide in the atmosphere was approximately 280 ppm for hundreds of years until it started to rise around 1800. In 1950, the carbon dioxide concentration in the atmosphere was approximately 311 ppm before rising more dramatically. [94] The recent value of approximately 420 ppm represents a 50 percent increase from the 280-ppm baseline.

Note that the vertical axis starts at 150 ppm, which can distort the perception of the change.

Figure 9-6 shows that the concentration of carbon dioxide in the atmosphere from 1800 to 2020, based on five-year average concentrations every fifth year, reached 300 ppm around 1910. Note that the vertical axis starts at 250 ppm.

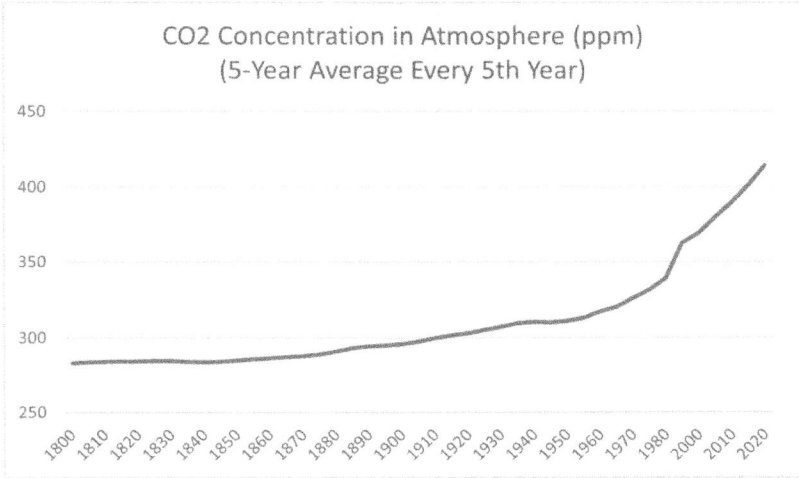

Figure 9-6 Carbon Dioxide Levels in the Atmosphere (1800-2020) [95] [96]

Direct air measurements provide more granularity and greater insight into the increase in the concentration of carbon dioxide in the atmosphere.

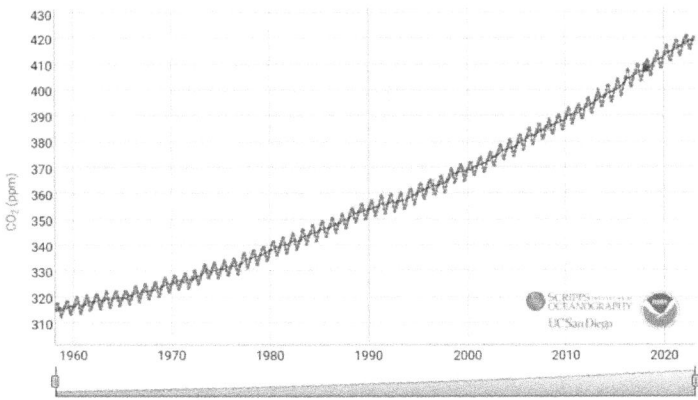

Figure 9-7 Mauna Loa Monthly Averages, 1960 to 2020 (NOAA) [97]

Figure 9-7 shows that the monthly average of the carbon dioxide concentration exhibits seasonal variations around its moving average. Shown are the annual peaks and valleys that are typically in the spring and fall, respectively, in the northern hemisphere, where most of the Earth's land mass outside of the tropics is located. Figure 9-8 shows the peaks and valleys with even more granularity. Notice that daily values can vary by up to approximately 1 ppm from one day to the next.

Mauna Loa Daily, Monthly and Weekly Averages for two years

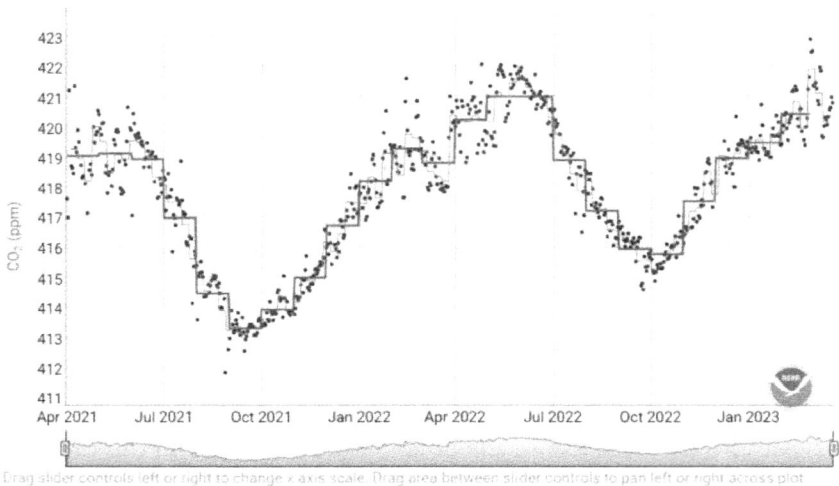

Figure 9-8 Mauna Loa Daily, Monthly and Weekly Averages 2021 to 2023 [98]

It might seem reasonable that human activity has caused and continues to cause carbon dioxide to accumulate in the atmosphere because, coincidentally, the carbon dioxide in the atmosphere started rising at the same time as humans started burning significant amounts of fossil fuels and generating large quantities of carbon dioxide.

However, **coincidence does not prove causality**. One way to prove causality is to examine an appropriate event on a granular basis. The ideal event would be to stop burning fossil fuels to cease producing carbon dioxide for a time and observe whether the carbon dioxide concentration in the atmosphere falls. Conversely, burning fossil fuels again should increase the concentration. This approach is not practical, because there is so much carbon dioxide in the atmosphere that it would take a long time to detect the changes. Further, this event would thrust the world back towards Level 1 and be devastatingly disruptive.

Carbon Dioxide Accumulation

The mass of carbon dioxide in the Earth's atmosphere is the carbon dioxide concentration in the atmosphere (420/1,000,000) times the mass of the atmosphere (5,150,000 billion metric tons, which is 5150 gigatons) times the ratio of the molecular weights of carbon dioxide and air (44/29), or approximately 3280 gigatons. The mass of carbon dioxide in the atmosphere in 1910 at a concentration of approximately 300 ppm was approximately 2340 gigatons (3280*300/420). Therefore, the atmosphere accumulated approximately 940 gigatons (3280-2340) of carbon dioxide between 1910 and 2023.

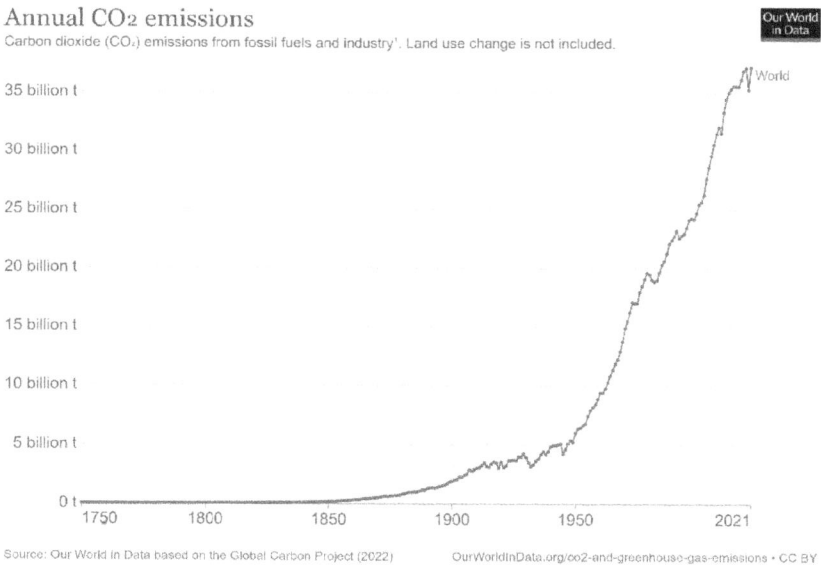

Annual CO₂ emissions
Carbon dioxide (CO₂) emissions from fossil fuels and industry¹. Land use change is not included.

Source: Our World in Data based on the Global Carbon Project (2022)   OurWorldInData.org/co2-and-greenhouse-gas-emissions • CC BY

1. **Fossil emissions**: Fossil emissions measure the quantity of carbon dioxide (CO₂) emitted from the burning of fossil fuels, and directly from industrial processes such as cement and steel production. Fossil CO₂ includes emissions from coal, oil, gas, flaring, cement, steel, and other industrial processes. Fossil emissions do not include land use change, deforestation, soils, or vegetation.

Figure 9-9 Annual Global Carbon Dioxide Emissions [99]

This same methodology was used in conjunction with emissions data in Figure 9-9 and carbon dioxide concentration data to develop Figure 9-10 comparing the amount of carbon dioxide emitted and absorbed. For reference, approximately 750 gigatons move through the natural carbon cycle each year. [100] Note that the vertical axis starts at a negative value to show when the data indicates that carbon dioxide was removed from the atmosphere.

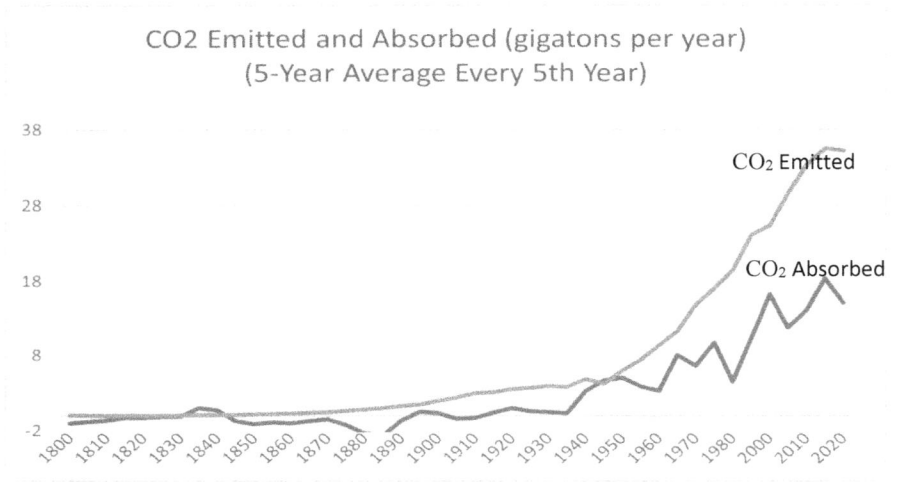

CO2 Emitted and Absorbed (gigatons per year)
(5-Year Average Every 5th Year)

Figure 9-10 Carbon Dioxide Emitted and Absorbed [101] [102] [103]

Under elevated carbon dioxide levels, most plant species show higher rates of photosynthesis and increased growth, [104] so the amount of carbon dioxide absorbed by the Earth's processes is higher at higher carbon dioxide concentrations. Stated differently, the amount of carbon dioxide absorbed by the Earth increases as its concentration increases and decreases when its concentration decreases. This can be seen by examining the carbon dioxide levels in Figure 9-6 in conjunction with the amount of carbon dioxide absorbed in Figure 9-10, understanding that part of the absorption is due to other processes.

Figure 9-11 shows that an average of 44 percent of the carbon dioxide emitted was absorbed by the Earth when the 5-year average carbon dioxide concentration in the atmosphere was between 313 ppm in 1955, and 415 ppm in 2020. It is not known what percentage of carbon dioxide emissions will be absorbed at higher carbon dioxide concentrations.

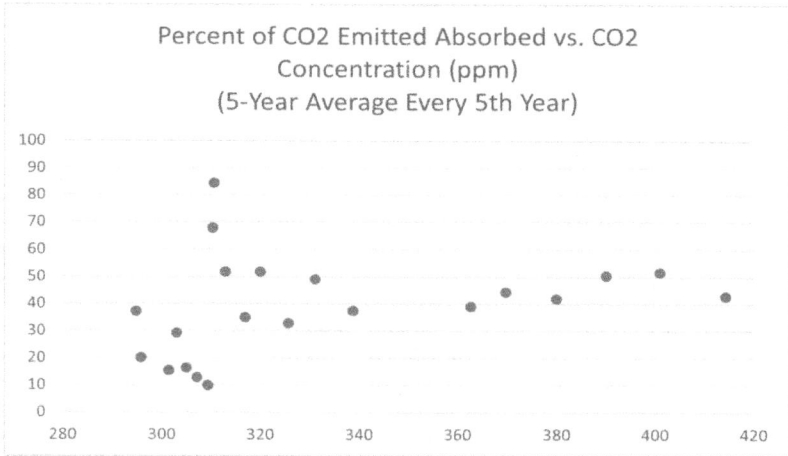

Figure 9-11 Carbon Dioxide Absorbed vs. Concentration [105] [106] [107]

Greening of the Earth

Higher concentrations of carbon dioxide in the atmosphere have a cooling effect on the land due to the increased efficiency of heat and water vapor transfer to the atmosphere. In addition, satellite images have observed increased green cover on land. [108] This implies that not only is more vegetation available to consume carbon dioxide, but it does so faster and provides a source of cooling.

Figure 9-10 suggests that if the upward trend reverses and the concentration of carbon dioxide in the atmosphere falls, then the Earth will retrace its trend by absorbing less carbon dioxide and reducing the amount of vegetation.

These findings help explain why the Earth absorbed much more carbon dioxide in the early 2020s as compared to 1910 (Figure 9-9) when the carbon dioxide concentration in the atmosphere started to increase (Figure 9-4, Figure 9-5, and Figure 9-6).

Carbon Dioxide Removal

In Figure 9-9, the amount of carbon dioxide that the Earth absorbed without materially affecting the carbon dioxide concentration was approximately 3 billion tons per year in 1910, when the concentration of carbon dioxide in the atmosphere was approximately 300 ppm. This suggests that reducing human carbon dioxide emissions to less than approximately 3 billion tons would eventually return the concentration back to approximately 300 ppm.

The carbon dioxide concentration in the atmosphere in 2023 was approximately 420 ppm, so each ppm represents approximately 7.8 gigatons (3280/420) of carbon dioxide. The rate at which carbon dioxide is added to the atmosphere is approximately 2.31 ppm per year. [109] Therefore, the atmosphere accumulates approximately 18 gigatons (2.31x7.8) of carbon dioxide annually, which is approximately half of the most recent carbon dioxide emissions in Figure 9-9.

This suggests that reducing human carbon dioxide emissions by approximately 18 gigatons per year would stop the accumulation of carbon dioxide in the atmosphere, because the amount emitted is approximately equal to the amount absorbed. Immediately reducing emissions to zero would enable the Earth to absorb approximately 18 gigatons of accumulated carbon dioxide each year and cause the concentration of carbon dioxide in the atmosphere to decrease.

Continuing to reduce the accumulated carbon dioxide by 18 gigatons per year would remove all the accumulated carbon dioxide in approximately 52 years (940/18). However, the amount of carbon dioxide that is removed will not remain at 18 gigatons per year, as suggested in Figure 9-10, because vegetation grows more slowly at lower concentrations, resulting in less removal of accumulated carbon dioxide. This effect will be partially offset if more vegetation is available to consume carbon dioxide than existed previously when the concentration increased. Nonetheless, it will take considerably longer than 52 years to return to a concentration of 300 ppm, and perhaps as long as a century. This is not surprising given that most of the carbon dioxide accumulated over approximately 100 years.

However, headwinds to the analysis in the previous two paragraphs include that the global population will increase until approximately 2100, additional people will increase their standard of living and consume more energy, new fossil fuel power plants are being built, and it will take decades to significantly reduce carbon dioxide emissions. Therefore, it will probably take more than a century to return the carbon dioxide concentration in the atmosphere to 300 ppm.

That said, there does not appear to be a need to achieve zero fossil fuel emissions, but rather maintain an emission level that stabilizes the carbon dioxide concentration of the atmosphere at a level that promotes vegetation growth without unduly affecting the environment.

Greenhouse Gases

A greenhouse gas can be characterized by its global warming potential (GWP), which is the multiple of the energy that it absorbs as compared to the amount of energy that would be absorbed by the same mass of carbon dioxide.

The GWPs of carbon dioxide, methane, and nitrous oxide are approximately 1, 83, and 273, respectively, [110] which means that methane and nitrous oxide are 83 and 273 times more potent than carbon dioxide on a mass basis. Methane is the major component in natural gas, whereas nitrous oxide is largely emitted from agriculture. Other greenhouse gases can be over 10,000 times more potent than carbon dioxide, and many such gases have been, or are being, phased out.

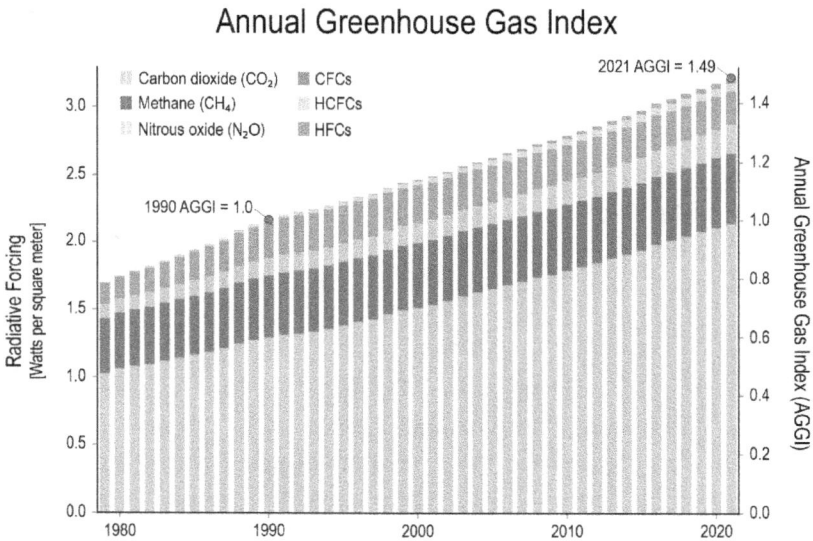

Figure 9-12 Annual Greenhouse Gas Index (AGGI) [111]

Figure 9-12 shows the annual greenhouse gas index (AGGI), depicting how much each greenhouse gas contributes to overall energy absorption in the atmosphere. Carbon dioxide accounts for approximately 67 percent (1/1.49) of the AGGI and is growing significantly. Methane accounts for approximately 15 percent of AGGI.

Significant sources of methane emissions were present naturally before humans made widespread use of natural gas, coal, and oil. Approximately 40 percent of these emissions originate from natural sources [112] and cannot be controlled. Therefore, approximately 9 percent of AGGI can be attributed to human activity, where agriculture and energy represent approximately 2.3 and 3.7 percent of AGGI, respectively. [113] Efforts to reduce methane emissions should continue, however major reductions in methane emissions will have little effect on the total AGGI, 67 percent of which is the result of carbon dioxide.

Therefore, the **primary focus of reducing total greenhouse gas emissions should be on carbon dioxide**, which is present in the products of combustion emitted by burning fossil fuels and has the largest effect on the environment. However, methane emissions attributable to human activities should also be reduced as much as practical.

Carbon dioxide can remain in the atmosphere for an indeterminate amount of time. In contrast, methane and nitrous oxide have lifetimes of approximately 12 and 109 years, respectively. [114] This means that the effects of today's methane and nitrous oxide emissions will linger in the atmosphere for approximately 12 and 109 years, respectively. Other greenhouse gases can linger for as long as 10,000 years or more.

COVID-19

The early 2020 worldwide slowdown in the initial throes of COVID-19 could provide granular data to show the effect of human activity on the concentration of carbon dioxide in the atmosphere. As illustrated in Figure 9-9, carbon dioxide emissions were 37.08 and 35.26 billion metric tons from fossil fuels and industry in 2019 and 2020, respectively. This is a reduction of approximately 1.8 billion metric tons of carbon dioxide.

The 1.8-billion-ton reduction in the generation of carbon dioxide in 2020 is approximately 10 percent of the 2020 atmospheric accumulation (1.8/18) and corresponds to approximately 10 percent of the 2.31 ppm annual increase, or approximately 0.23 ppm. The initial reduction in early 2020 may have been more severe than the annualized 0.23 ppm reduction indicates. However, examination of Figure 9-8 reveals that measurements can vary by as much as 0.75 ppm from one day to the next. As significant and disruptive as the COVID-19 slowdown was, researchers could not identify a significant carbon dioxide concentration reduction resulting from the COVID-19 slowdown.

By way of example, Figure 9-13 displays this methodology using contemporaneous measurements that are unrelated to global warming to identify a dramatic drop in nitrogen dioxide concentration before and during the COVID-19 slowdown. In contrast, the accumulation of carbon dioxide was so large that daily measurements could not detect any significant change.

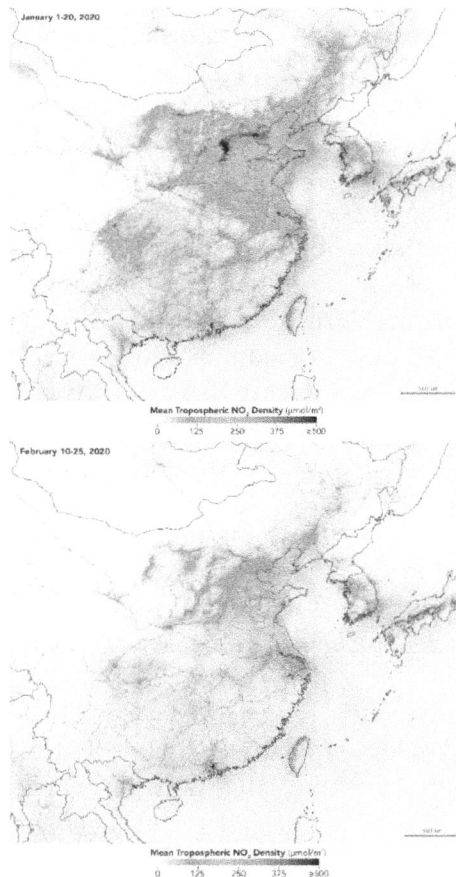

Figure 9-13 NO$_2$ Concentrations in China Before and During COVID-19 Shutdown [115]

Chapter 10: The Effects of Greenhouse Gas

You can't stop progress, but you can help decide what is progress and what isn't. --- Ashleigh Brilliant

High school physics was offered primarily to science and engineering students in their senior year in my high school class of approximately 500 students. There was only one physics teacher who taught four or five classes of approximately 16 students each, so approximately 15 percent of students studied physics. Therefore, most students fell into the other 85 percent that did not have this experience.

Some years later, my daughter approached me during the first week of school and complained about being totally frustrated with physics. She said that she had memorized the motion equations but did not understand what to do with them. After some discussion, she confided that her academic success was primarily predicated on her ability to memorize material and do well on tests. Physics had her stifled. My advice to her was to work on problems in her textbook. But she would have none of that.

We discussed her progress a couple days later, only to discover that she was even more frustrated. Physics is different from other subjects in the sense that a person can know the equations well and not have the slightest idea what to do with them. Knowing what to do comes not from memorizing but rather from solving problems using equations. In other words, working on problems is practicing, and practicing helps us learn.

By the next week, she had admitted to herself that memorizing equations was futile and had started working on problems. She ultimately learned how to solve physics problems, earned an A in physics, and was introduced to another way to learn.

---xxx---

There is a plethora of information out there, but most of it is irrelevant to the problem at hand. There are things we know that we know, and things that we know that we do not know. For example, I know that I can speak English, but I also know that I cannot speak virtually any other language. There are things that we don't know that we know. For example, some years ago, I tuned out of a conversation about sports, which I do not follow, until I remembered that I was knowledgeable about the specific events being discussed.

The physics student did not know what she did not know. In her case, she did not know that some subjects, such as physics, are learned in different ways. Similarly, some skills are learned differently. Knowing how to learn about a specific subject or what data to examine as part of an investigation is often more of an art than a science.

Spoiler Alert: This chapter presents an abundance of seemingly disparate information that has been associated with global warming in one way or another. Be aware that some of this information is relevant to global warming, and some is not.

Greenhouse Effect

Earth would be a cold, barren place if not for the greenhouse effect, where trapped energy enables our atmosphere to maintain temperatures that support life as we know it.

The Earth receives radiation from the Sun, some of which is absorbed in the Earth's atmosphere and by the Earth itself. However, a portion of that energy is radiated back into Space, as can be seen in Figure 10-1. Therefore, the amount of energy that remains on Earth is equal to the amount of energy that arrives from the Sun minus the energy that is radiated by Earth into Space. Both energy streams are filtered by various components in the atmosphere.

The upper panel in Figure 10-1 shows energy traveling from the Sun to the Earth in the form of ultraviolet, visible, and infrared radiation distributed among various wavelengths as defined by the 5525K line in the left half of the graph. The solid parabola under the 5525K line shows the amount of energy that reaches Earth at different wavelengths after some energy is absorbed by the major components, as shown in the small lower panels.

The 210-310K curves on the right show that the energy that radiates from Earth at different Earth temperatures is overwhelmingly infrared energy. The solid parabola-like areas under the 210-310K curves show the amount of energy that reaches Space at different wavelengths after some energy is absorbed by the major components. The Earth would be much colder if the heat from these components was not absorbed in the atmosphere.

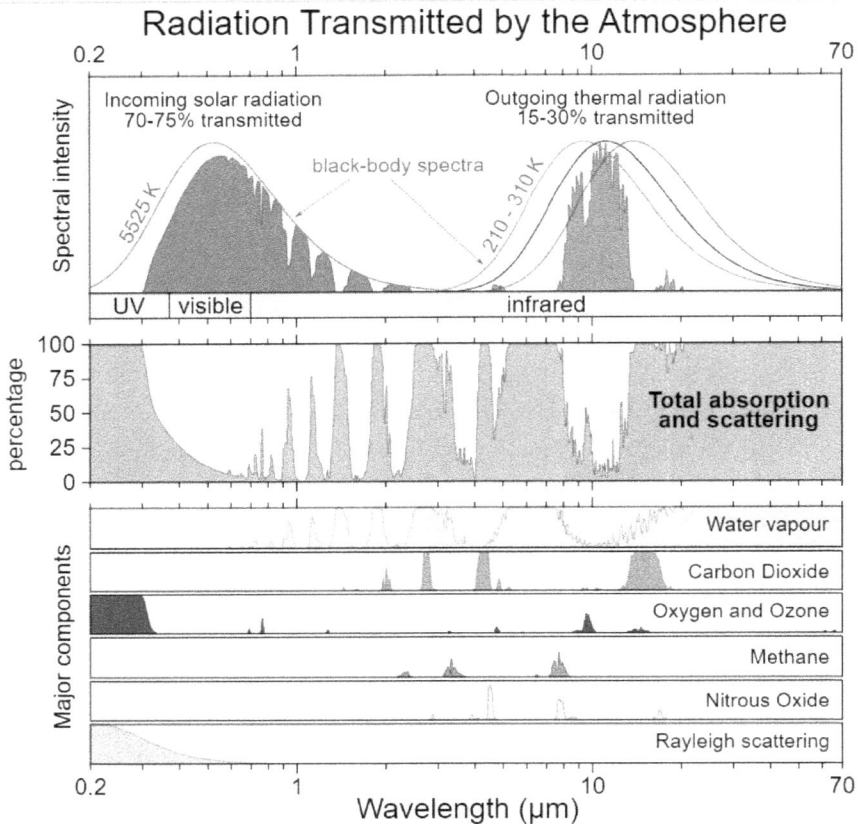

Figure 10-1 Radiation Transmitted by the Atmosphere [116]

By observation of Figure 10-1, energy that arrives from the Sun has relatively short wavelengths, while energy that leaves the Earth has relatively long wavelengths. The atmosphere filters both energy entering and leaving Earth. The percentage of energy that passes through the atmosphere at various wavelengths after being filtered by the major components is shown in the middle panel of Figure 10-1.

Carbon dioxide and water are considered greenhouse gases because they absorb energy from Earth that would otherwise be radiated into Space. Figure 10-2 shows that water vapor has the most prominent influence because it almost completely blocks energy over wide ranges of wavelengths when there are clouds.

106

Carbon dioxide is shown to block energy leaving Earth at several wavelengths, regardless of whether clouds are present or not.

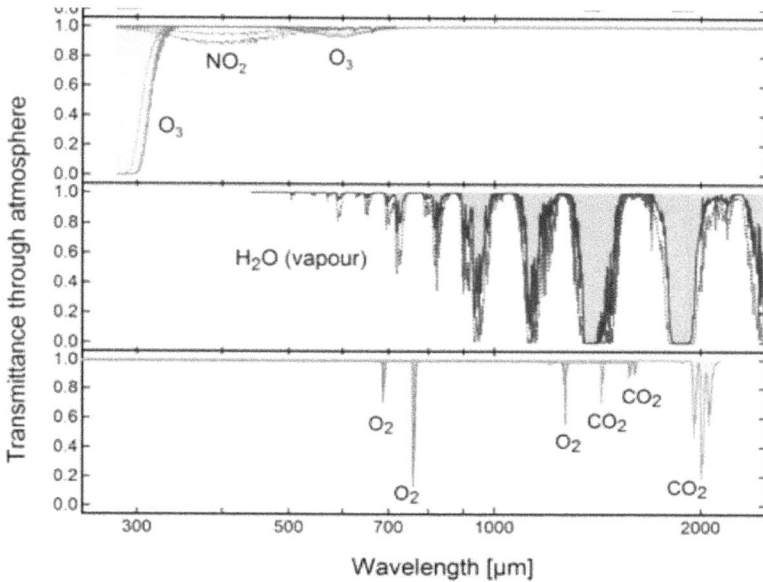

Figure 10-2 Sources of Absorption [117]

Carbon dioxide is inert and can remain in the atmosphere for long periods of time before being consumed by photosynthesis, absorbed by water, or otherwise removed. An increase in the concentration of carbon dioxide in the atmosphere causes more heat to be retained on Earth than would normally occur at its pre-1800 concentration.

Water vapor is the largest greenhouse gas, but it only remains in the atmosphere for a few days. The amount of water vapor in the atmosphere varies with location and weather conditions. For example, hot, dry deserts contain precious little water vapor, while the air in tropical rainforests can be extremely humid and contain almost the maximum amount of water vapor possible.

Surface Temperature Data

The average global surface temperature is commonly used to describe the temperature of the Earth for climatic purposes.

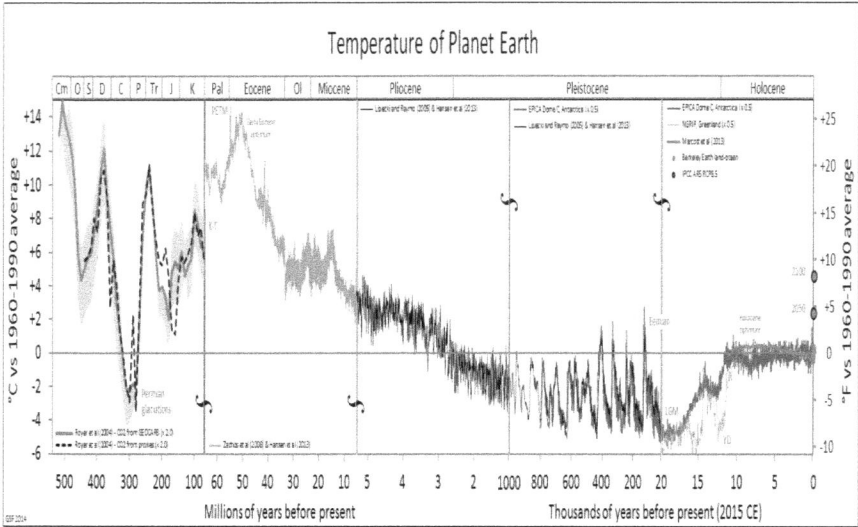

Figure 10-3 Temperature of Planet Earth [118]

Figure 10-3 shows the temperature of the Earth over the last 500 million years. Note that the horizontal axis is logarithmic, so what appears to have been abrupt changes millions of years ago took place over hundreds of thousands, or millions of years. The average temperature has been relatively stable for approximately the last 10,000 years. The 2050 and 2100 dots are projections that may or may not materialize.

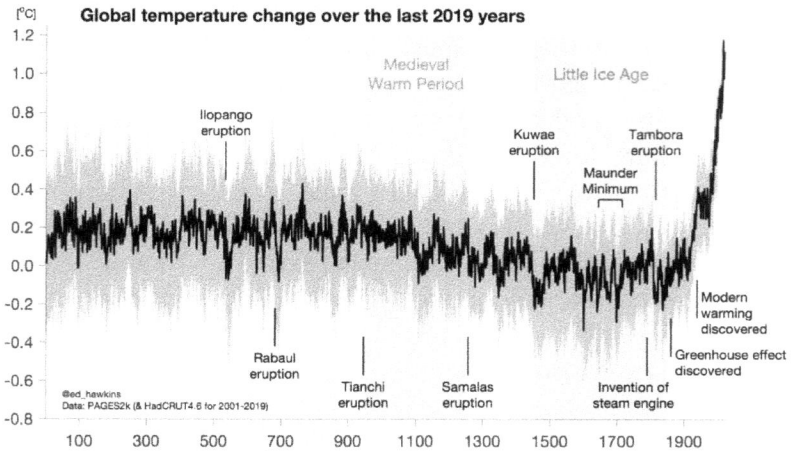

Figure 10-4 Global Temperature Change Over the Last 2019 Years [119]

Figure 10-4 shows the Earth's temperature with more granularity over the last 2000 years, with annotations showing significant events. By observation, the temperature of the Earth broke out of its previous pattern and started to rise abruptly in the early 1900s, stabilizing in the middle of the 1900s, and then continuing to rise.

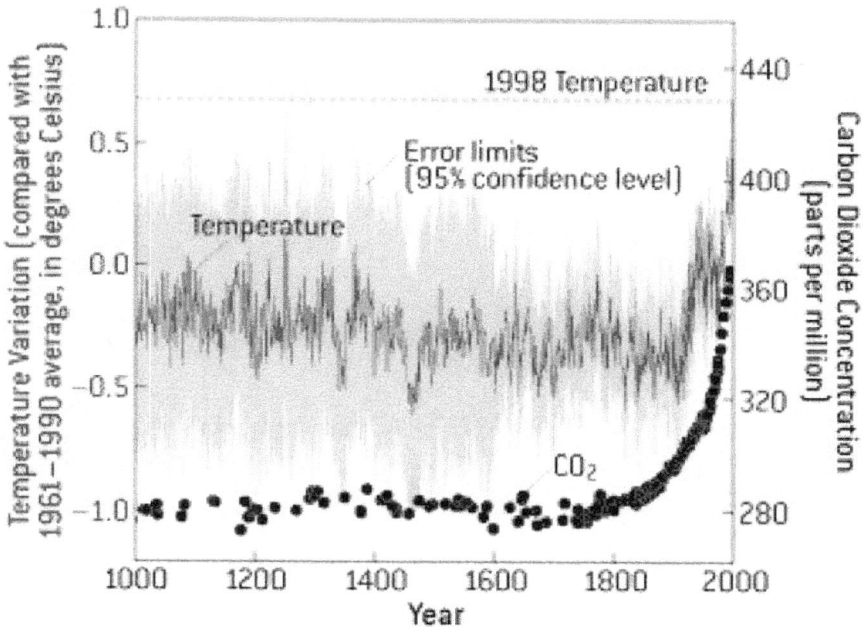

Figure 10-5 Temperature Variation and Carbon Dioxide Concentration [120]

Figure 10-5 provides more granularity and shows that the carbon dioxide concentration in the atmosphere started increasing around 1800. However, the temperature started to increase steadily in 1977, as shown in Figure 10-6.

Note that the left side of Figure 10-6 shows that the global average surface temperature decreased, and in the 1970s, there were concerns about global cooling. [121] Importantly, the change in carbon dioxide concentration around 1950 precedes the temperature change that occurred around 1977. Therefore, this **data suggests that, if the relationship between the carbon dioxide concentration and temperature on Earth is causal, the cause is the carbon dioxide concentration.**

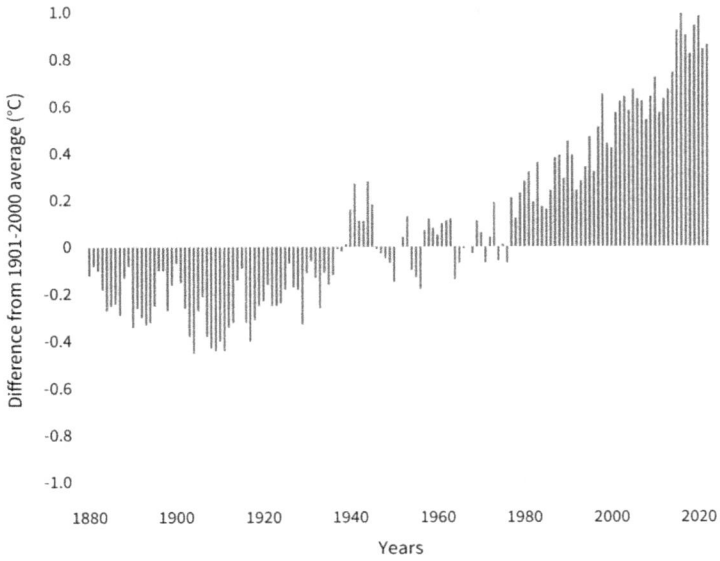

Figure 10-6 Global Average Surface Temperature [122]

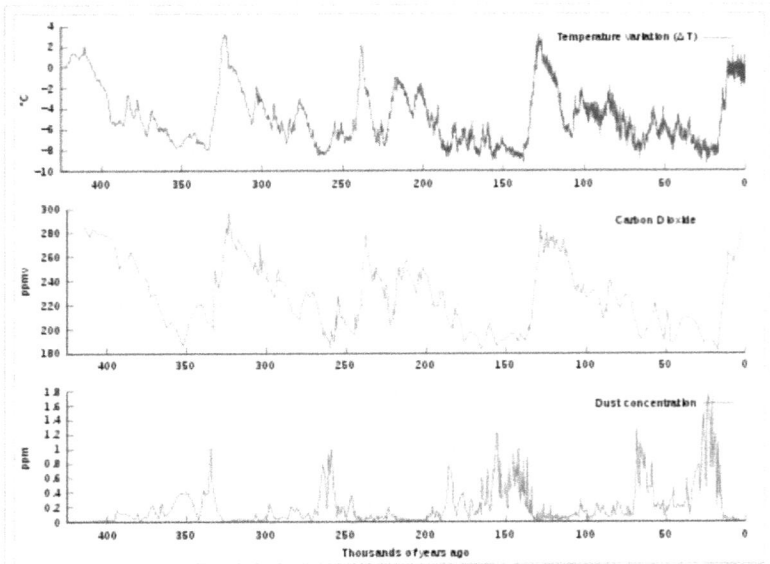

400,000 years of ice core data

Figure 10-7 Temperature, Carbon Dioxide, and Dust Concentration
(400,000 years) [123]

Examination of the temperature and carbon dioxide curves in Figure 10-7 leads one to conclude that there is a causal relationship between these two variables because they are remarkably similar and almost identical in virtually every rise and fall over a long period of time. **This observation, in conjunction with the data in Figure 10-5 and Figure 10-6, suggests that carbon dioxide in the atmosphere caused the temperature of the Earth to change.**

Heat Waves

Heat waves are prolonged periods of hot weather. Figure 10-8 shows that the frequency, duration, and intensity of heat waves have been increasing since the 1960s. However, the data is from metropolitan areas that are *urban heat islands* whose temperatures can be higher than surrounding rural areas because vegetation is often lost, and more surfaces are paved or covered with buildings. 124

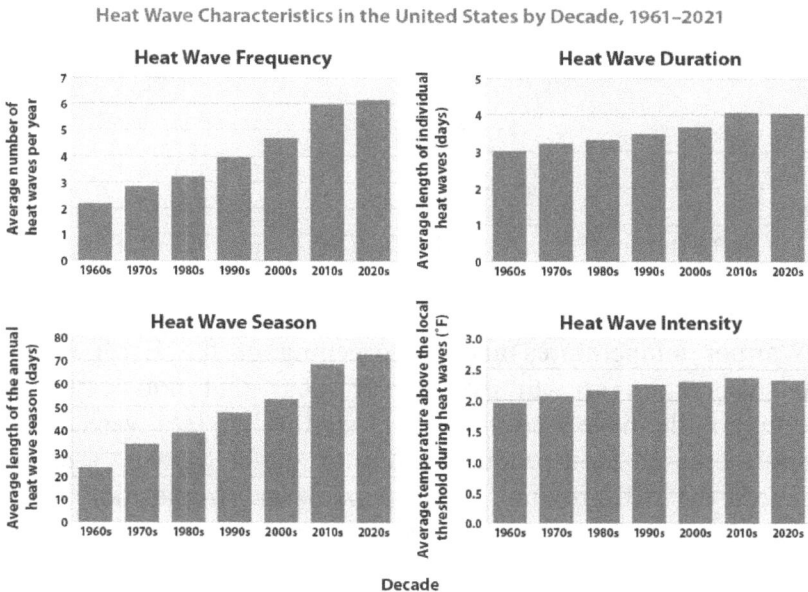

Heat Wave Characteristics in the United States by Decade, 1961–2021

Data source: NOAA (National Oceanic and Atmospheric Administration). 2022. Heat stress datasets and documentation. Provided to EPA by NOAA in February 2022.

For more information, visit U.S. EPA's "Climate Change Indicators in the United States" at www.epa.gov/climate-indicators.

Figure 10-8 Heat Wave Characteristics in the United States (1961-2021) [125]

Figure 10-9 shows that the annual heat wave index is consistent with Figure 10-8 between 1961 and 2021. However, the annual heat index was somewhat higher between 1895 and 1960, especially in the 1930s. This does not support the premise that the Earth is warming but rather shows how utilizing incomplete data can lead to an erroneous conclusion.

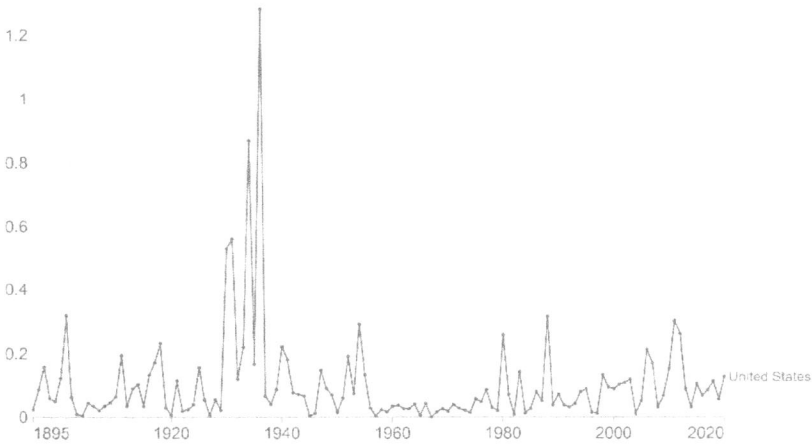

Figure 10-9 Annual Heat Wave Index in the United States (1895-2021) [126]

Agriculture

Warmer temperatures and the greening of the Earth result in faster vegetation growth over wider areas that not only absorb carbon dioxide, have a cooling effect, and utilize less water but also enable increased food production during longer growing seasons to help feed the still-growing global population. Ample food supplies help provide more human security and potentially promote prosperity.

Peace and Prosperity

Extensive historical research in China reveals a close and positive relationship between a warmer climate and peace and prosperity, and between a cooler climate and war and poverty. [127]

Sea Level Rise

Increasing temperatures can cause ice in the polar ice caps to melt and raise sea levels around the world. The following graphs, with increasing granularity, show how sea levels have changed over the last 24,000 years.

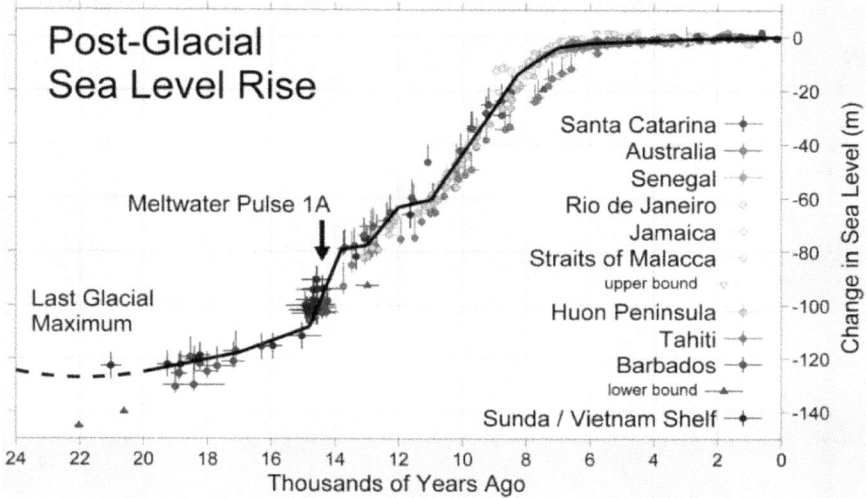

Figure 10-10 Post-Glacial Sea Level Rise [128]

Figure 10-10 shows that the sea level rose 120 meters over a period of thousands of years before stabilizing approximately 6000 years ago. It shows that the sea level rose at a rate of approximately 4000 centimeters over a period of 4000 years, or one centimeter annually. This substantial change occurred well before humans burned fossil fuels.

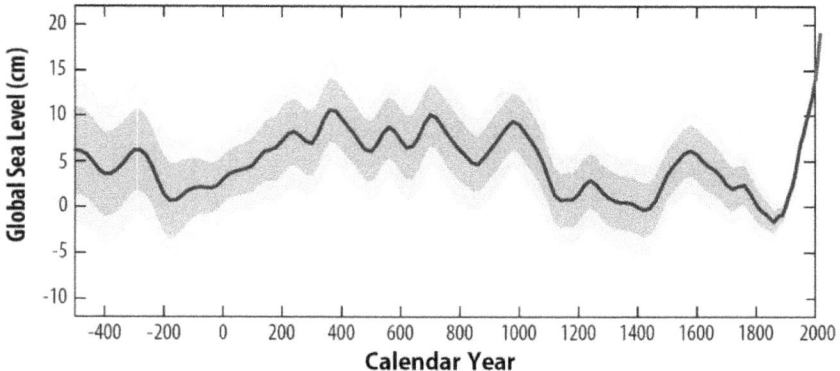

Figure 10-11 Global Sea Level [129]

113

Figure 10-11 shows that the sea level has varied somewhat over the last 2500 years.

Figure 10-12 shows that the sea level rose approximately 4 centimeters in the 30 years between 1990 and 2020, or 1.33 centimeters annually. A sea level change that is faster than 10 percent of a major change, such as the one cited above, is not insignificant, and especially so when it starts at the same time the Earth is warming.

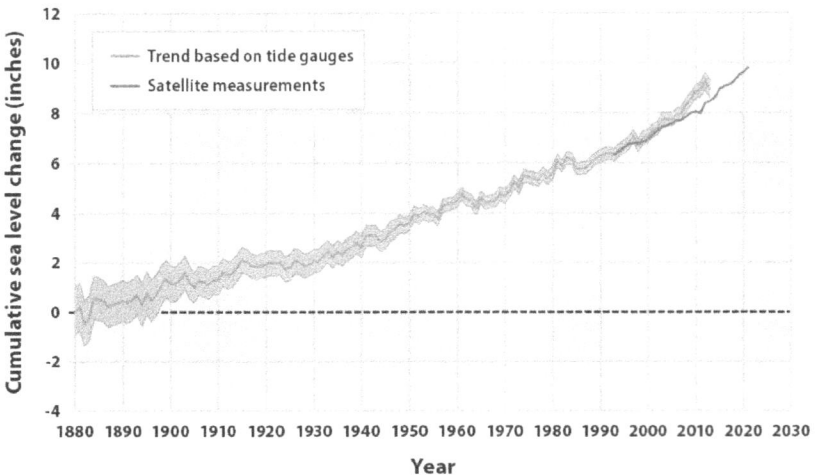

Figure 10-12 Global Absolute Sea Level Change, 1880-2021 [130]

Coastal Flooding

The sea level rose above its 2000-year range around 1950 and accelerated, which could be a coincidence, but it may be causal if the polar ice caps are melting at the same time. Extreme melting could potentially inundate the coasts and coastal cities to the extent that mitigation may be necessary.

Figure 10-13 shows that coastal flooding in the United States has increased since the 1950s.

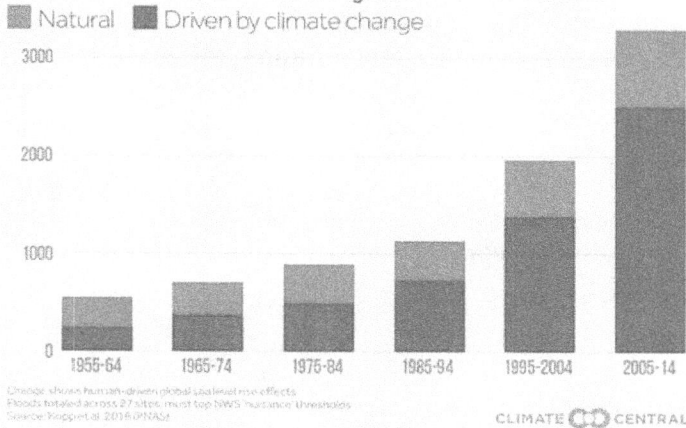

Coastal Flood Days in the U.S.

■ Natural ■ Driven by climate change

Figure 10-13 Coastal Flood Days in the United States [131]

Ocean Acidity

The ocean absorbs about 30 percent of the carbon dioxide that is released into the atmosphere. [132] The chemical reactions that occur increase the acidity of the ocean. Carbonate ions become less abundant and make life more difficult for organisms that build calcium carbonate structures, such as shellfish, corals, and calcareous plankton. These changes in ocean chemistry can affect other species as well.

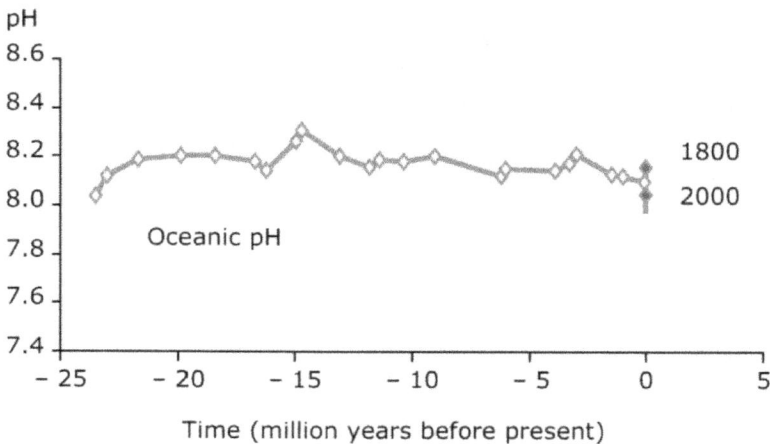

Figure 10-14 Ocean Acidity Over the Past 25 Million Years (without predictions) [133]

115

Figure 10-14 and Figure 10-15 show historical data for ocean acidity with increasing granularity. Ocean acidity can be seen to have been relatively stable for millions of years but decreased substantially in the late 1900s and early 2000s as the carbon dioxide concentration in the atmosphere increased. This relationship could be coincidental or causal.

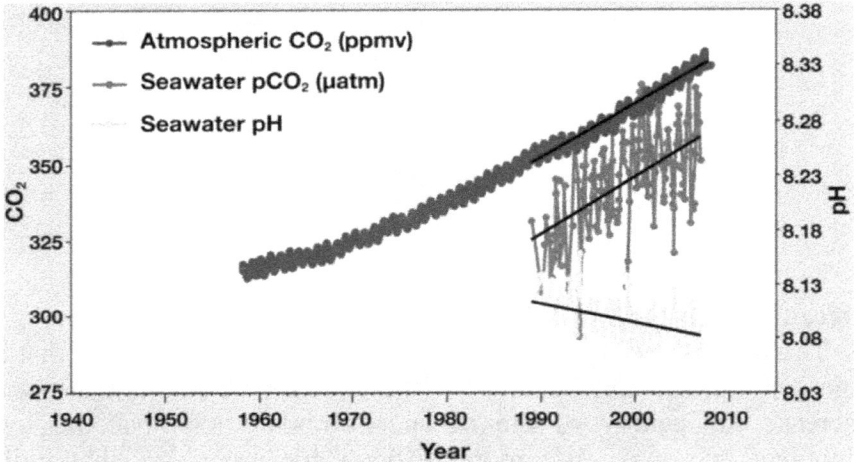

Figure 10-15 Ocean Acidification [134]

Snow-Covered Area

The snow-covered area in North America is shown in Figure 10-16 where notes in the figure indicate that the snow cover was approximately two percent lower during the last decade than in the first decade. In addition, the average snow cover season in the United States became shorter by nearly two weeks. This was due to the last day of snow shifting earlier by 19 days since 1972, while the first date of snow cover remained relatively unchanged. Overall, these observations are consistent with an increase in the temperature of the Earth.

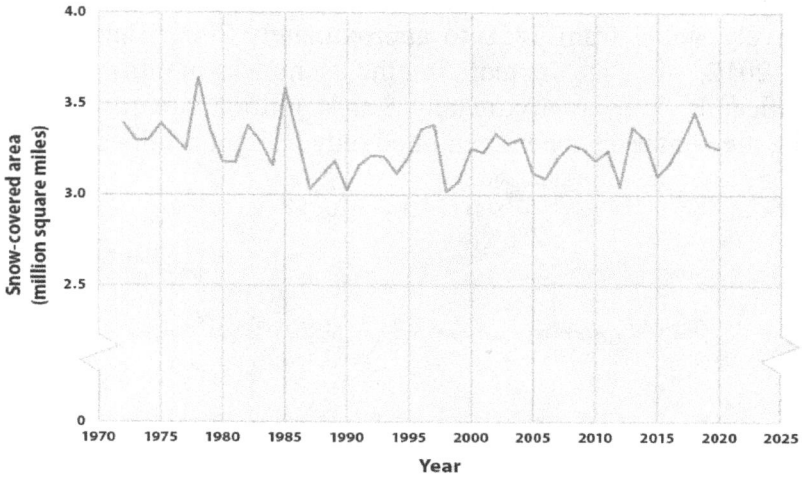

Figure 10-16 Snow-Covered Area in North America [135]

Polar Ice

Figure 10-17 shows that the Arctic Sea ice extent migrated over time from the top solid line (1980) to the dashed line (2012) to the solid middle line (2022), which is above the minimum reached in 2012, but below the normal range.

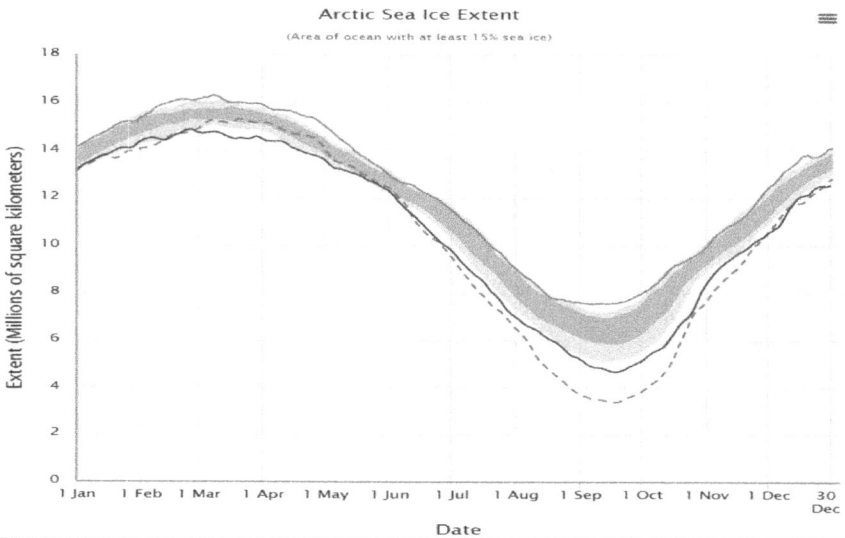

Figure 10-17 Arctic Sea Ice Extent [136]

117

Figure 10-18 shows that the ice extent in the winter has remained relatively stable from 1850 to approximately 2000. Between 2000 and 2012, the ice extent in the summer months decreased significantly from approximately 8 to 4 million square kilometers, while the winter ice extent exhibited only a slight decline.

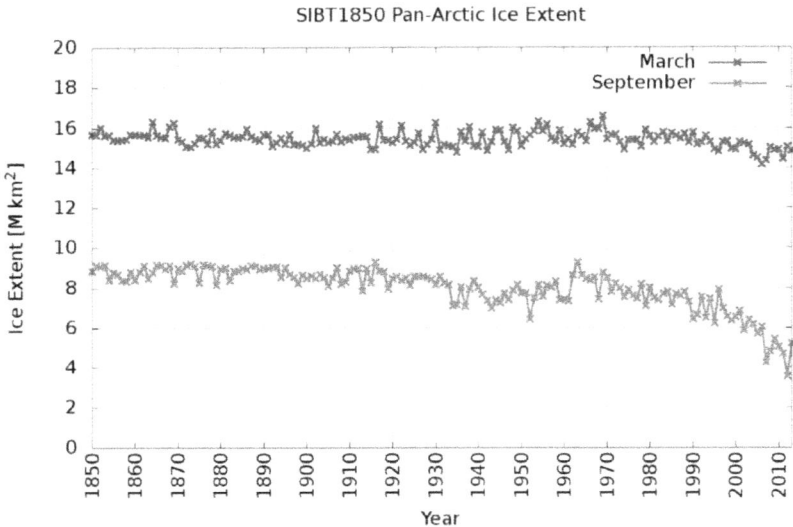

Figure 10-18 1850 Arctic Sea Ice Extent [137]

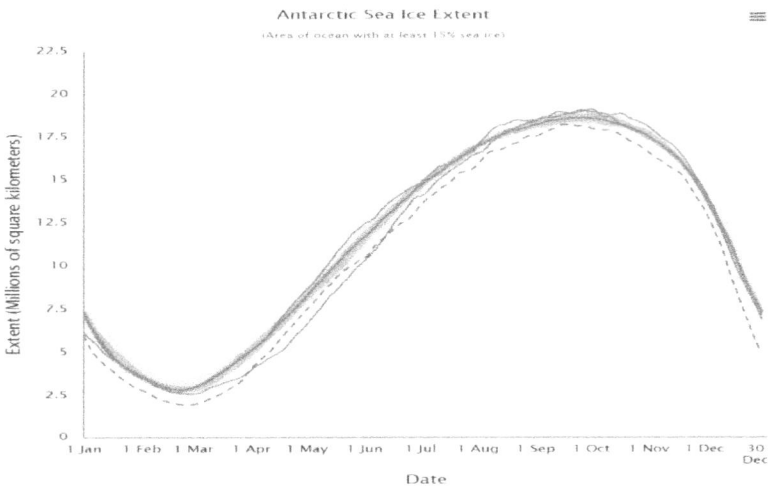

Figure 10-19 Antarctic Sea Ice Extent [138]

118

Figure 10-19 shows the Antarctic Sea ice extent for 1980 and 2000 (solid lines). The dashed line (2022) is the record minimum.

The data on the ice extent graphs is largely consistent with an increase in the Earth's temperature.

Polar Bear Population

There is concern that the polar bear population could be reduced because there is less Arctic Sea ice from which polar bears can hunt for seals. This could create local issues for the polar bears. However, Figure 10-17 indicates that the polar bears still have millions of square kilometers of Arctic Sea ice in which to hunt.

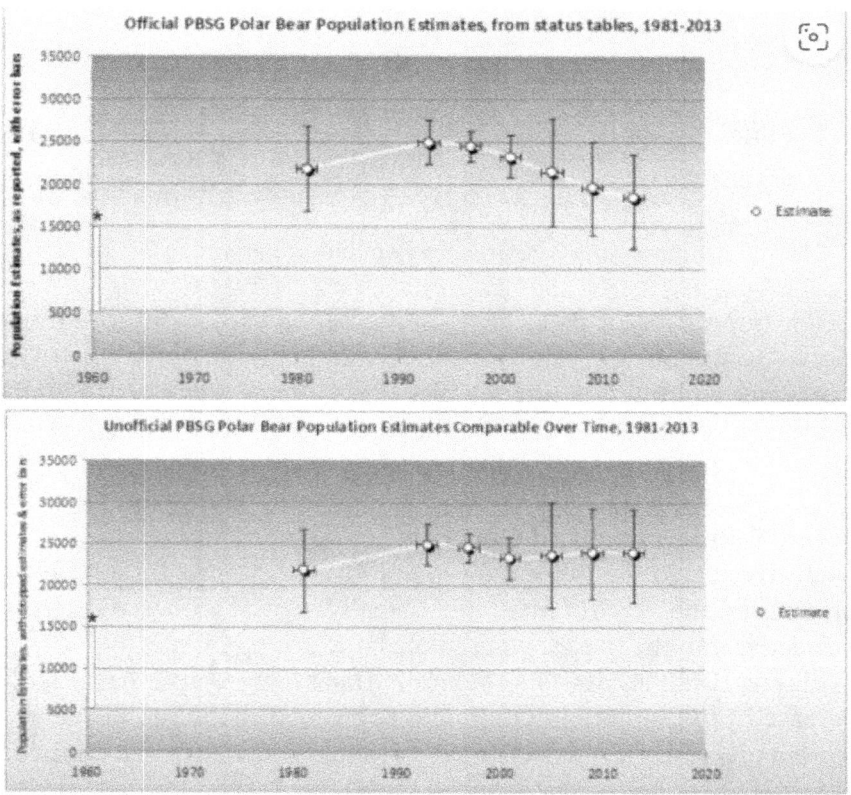

Figure 1. Upper graph uses totals reported in PBSG status tables, with min/max; Lower graph uses the same figures, but adds back in the so-called "inaccurate" estimates dropped 2005-2013. The 1960 figure * is a **ballpark estimate**

Figure 10-20 PBSG Polar Bear Population Estimates [139]

The upper part of Figure 10-20 is an estimate that shows a decline in the polar bear population, whereas the lower part adds back "inaccurate" estimates and shows the population to be stable. Figure 10-21 shows a later estimate indicating that the polar bear population is larger than previously determined.

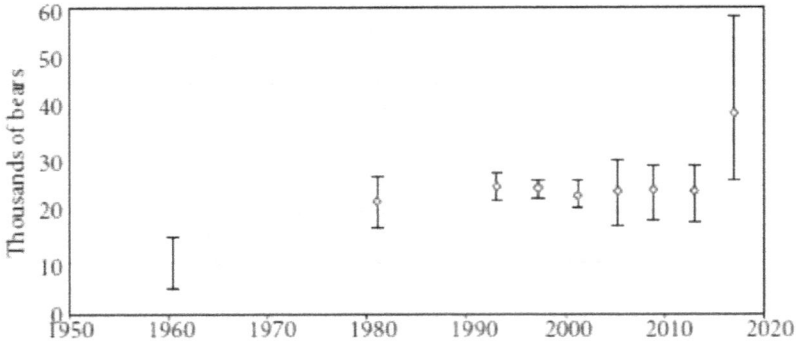

Global polar bear population size estimates to 2018. From Chapter 10 of *The Polar Bear Catastrophe That Never Happened* (Crockford 2019).

Figure 10-21 Polar Bear Population Estimate [140]

The population of polar bears in the Arctic is difficult to estimate due to the remote and inhospitable locations in which they live. The conflicting information implies that the available data is not reliable, and that no conclusion can be drawn as to whether the increase in the Earth's temperature is affecting the polar bear population.

Ocean Temperature and Currents

Heat rejected to the environment from burning fossil fuels is rejected to the environment but usually flows, either directly or indirectly, into the ocean, where it has a warming effect.

Figure 10-22 shows that the sea surface temperature has steadily increased by approximately 1 degree Celsius (1.8 degrees Fahrenheit) since around 1910. Interestingly, the increase coincides with the increase in the concentration of carbon dioxide in the atmosphere that began around 1910. Reliable data could not be located to determine whether this temperature rise is outside of normal historical values.

Nonetheless, this data begs the question as to whether the sea surface temperature rise is due to the thermal footprint of burning fossil fuels, or something else.

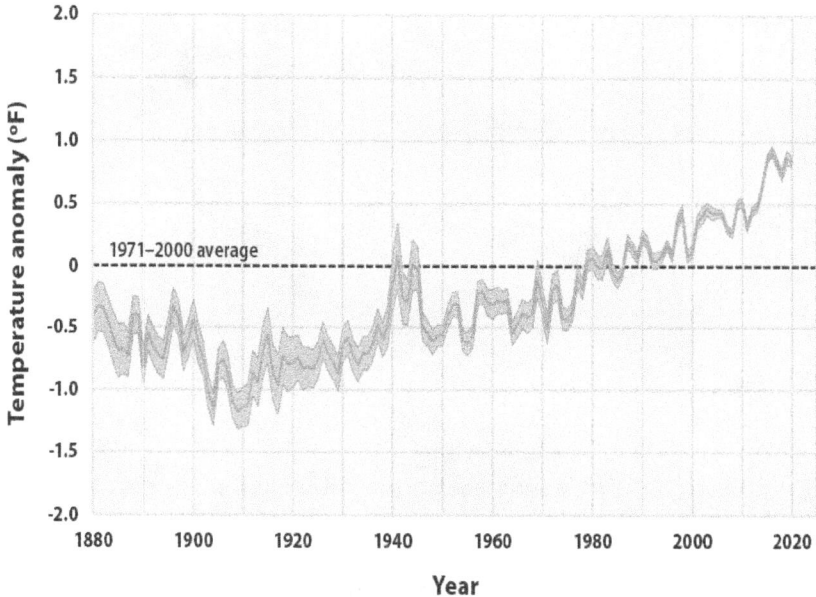

Figure 10-22 Average Global Sea Surface Temperature (1880-2020) [141]

Per Figure 10-23, approximately half of the CO_2 emissions since 1750 occurred during the 30-year period between 1990 and 2020. Therefore, approximately half of all fossil energy burned occurred during these same years.

By observation, Figure 10-24 shows that the average global fossil fuel consumption between 1990 and 2020 was approximately 110 terawatt hours (0.11 quadrillion watt hours).

Cumulative CO₂ emissions by world region

Cumulative carbon dioxide (CO₂) emissions by region from the year 1750 onwards. This measures CO₂ emissions from fossil fuels and industry only – land use change is not included.

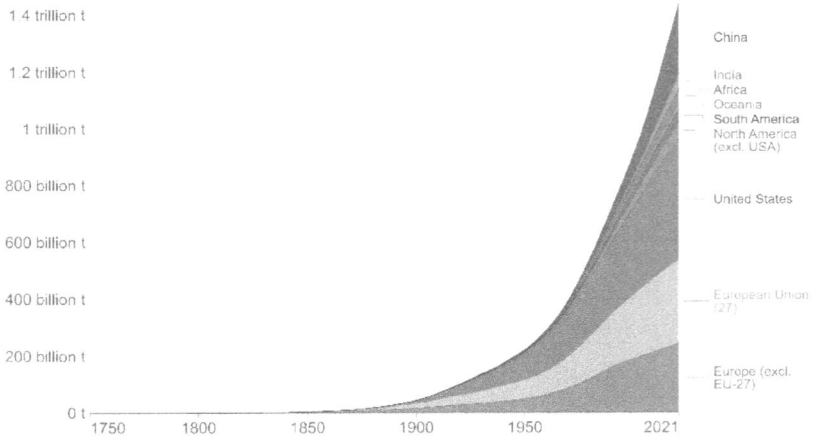

1. **Fossil emissions**: Fossil emissions measure the quantity of carbon dioxide (CO₂) emitted from the burning of fossil fuels, and directly from industrial processes such as cement and steel production. Fossil CO₂ includes emissions from coal, oil, gas, flaring, cement, steel, and other industrial processes. Fossil emissions do not include land use change, deforestation, soils, or vegetation.

Figure 10-23 Cumulative Carbon Dioxide Emissions [142]

Global fossil fuel consumption

Global primary energy consumption by fossil fuel source, measured in terawatt-hours (TWh).

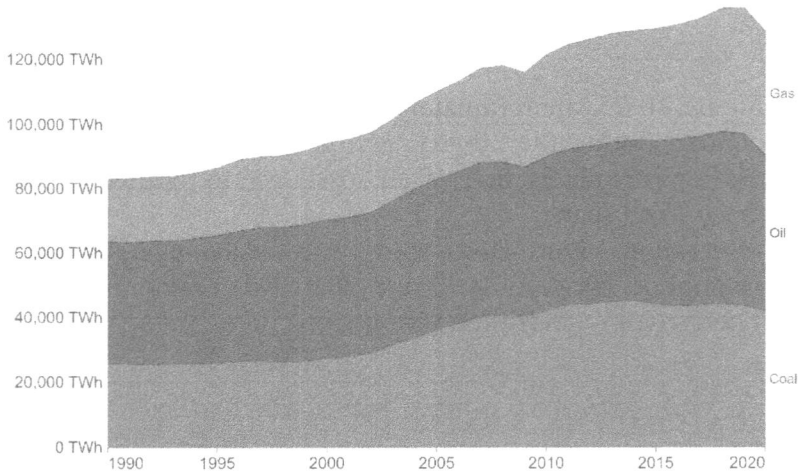

Figure 10-24 Global Fossil Fuel Consumption (1990-2000) [143]

The amount of heat released from fossil fuel since 1750 is approximately 22.5 quadrillion BTU (2*30 *0.11*3.413), where 1 kilowatt-hour is equivalent to 3.413 BTU. The surface area of water on Earth is approximately 140 million square miles times 28 million square feet per square mile, or 3.9 quadrillion square feet. The approximate weight of the first meter (3.28 feet) of water is its surface area (3.9 quadrillion square feet) times the density of seawater (64 pounds per cubic foot) times the depth of 3.28 feet, which is approximately 819 quadrillion pounds.

It takes approximately 1 BTU to raise the temperature of one pound of water (or sea water) by one degree Fahrenheit. If all the heat released from fossil fuel since 1750 were absorbed by the top meter of water, its temperature would rise approximately 0.015 degrees Celsius {22.5/(819*1.8)}, which is negligible in comparison with data showing an approximate 1 degree Celsius rise in the sea surface temperature. Therefore, the thermal footprint of burning fossil fuels did not cause the sea surface temperature to increase. However, a significant rise in water temperature can occur near thermal sources.

The rise in sea water temperature, which coincidentally started when the concentration of carbon dioxide in the atmosphere started increasing, should not be ignored because it is consistent with the effects of greenhouse gases. However, this does not prove causality, because the temperature rise could be the result of another environmental process.

Of note is that the carbon dioxide concentration started rising slowly, and then increased more rapidly. However, Figure 10-22 shows that the rise in sea surface temperature is essentially linear. One might expect the sea surface temperature rise to accelerate as the increasing carbon dioxide concentration trapped more heat on Earth. However, the data shows that this did not occur.

Figure 10-25 depicts the so-called ocean conveyor belt that circulates sea water and transports nutrients throughout the Earth's oceans.

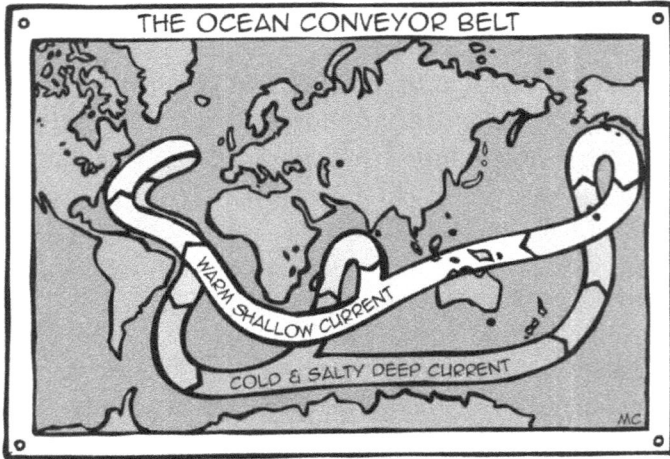

Figure 10-25 Ocean Conveyor Belt [144]

There is a concern that the warming Earth and melting ice might affect these flows and adversely affect the environment.

Precipitation

Figure 10-26 shows annual precipitation anomalies worldwide. The precipitation trend of only +1.00 inches per century and the presence of both positive and negative anomalies suggest that there is no causal link between the increasing concentration of carbon dioxide in the atmosphere and precipitation.

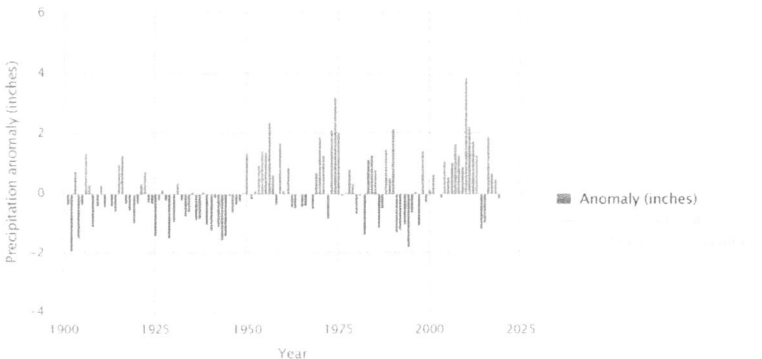

Exhibit 6. Annual precipitation anomalies worldwide, 1901–2019

Figure 10-26 Annual Precipitation Anomalies Worldwide, 1901 to 2019 [145]

124

Tornadoes, Storms, and Hurricanes

Reporting on virtually every major tornado, storm, and hurricane predictably brings claims that it was caused by climate change or global warming. Figure 10-27 shows that the number of tornadoes in the United States was stable for approximately 40 years before increasing abruptly in the 1990s. This, along with the significant decline in the number of tornadoes between 2000 and 2020 suggests that there is no causal link between the increasing concentration of carbon dioxide in the atmosphere and tornadoes.

However, the sudden tripling of tornadoes in the 1990s could have occurred naturally or perhaps is the result of improved tornado detection methods that detect tornadoes that were previously missed. If the latter suggestion is true, usefulness of data prior to the 1990s is dramatically reduced.

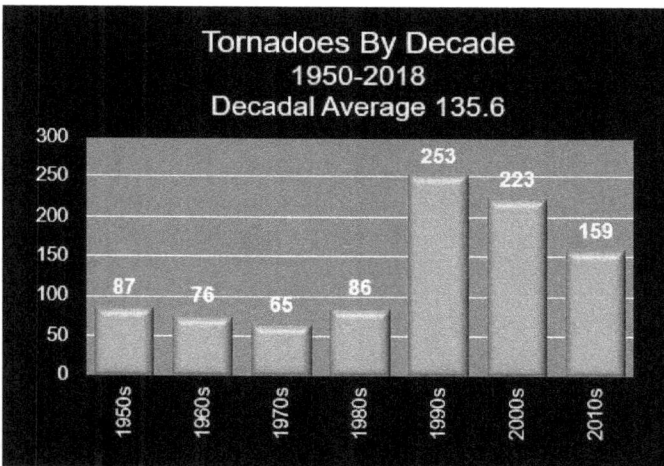

Figure 10-27 Tornadoes by Decade, 1950-2018 [146]

Figure 10-28 shows the number of storms and hurricanes in the United States. It has been adjusted to add storms that were missed due to technology issues that caused incomplete reporting at the time. The gradual increases and decreases in the number of storms and hurricanes suggest that there is no causal link between the effects of the increasing concentration of carbon dioxide in the atmosphere and these events.

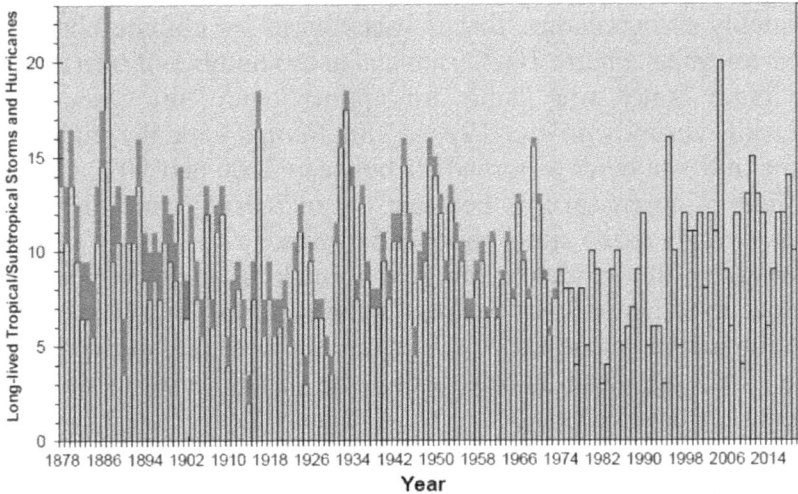

Adjusted Long-lived Tropical/Subtropical Storms and Hurricanes
Adding "Missed" Systems - Duration greater than 2.0 days - 1878 to 2020

Figure 10-28 Storms and Hurricanes, 1878 to 2020 (adjusted) [147]

Wildfires

Figure 10-29 shows that the frequency of wildfires reported by the National Interagency Fire Center in the United States remained relatively stable between 1985 and 2022, but with a downward trend after 2006.

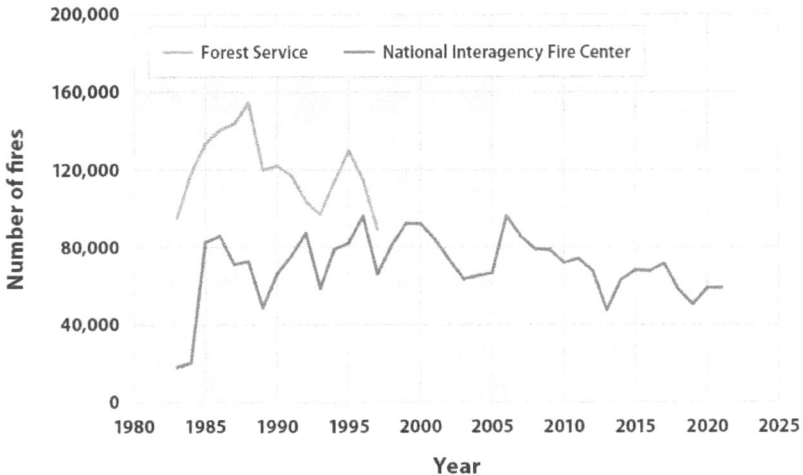

Figure 10-29 Wildfire Frequency in the United States, 1983–2021 [148]

126

The area burned by wildfires reported by the National Interagency Fire Center in the United States increased between 1985 and 2022, is shown in Figure 10-30. This could be a direct result of poor forest management that creates conditions for more destructive forest fires. [149]

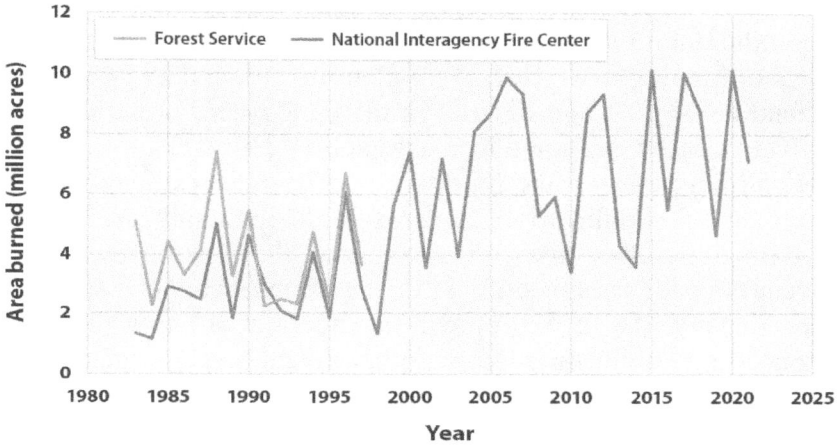

Figure 10-30 Wildfire Area Burned in the United States, 1983–2021 [150]

Figure 10-31 shows that the annual global wildfire carbon emissions have been trending downward between 2003 and 2022.

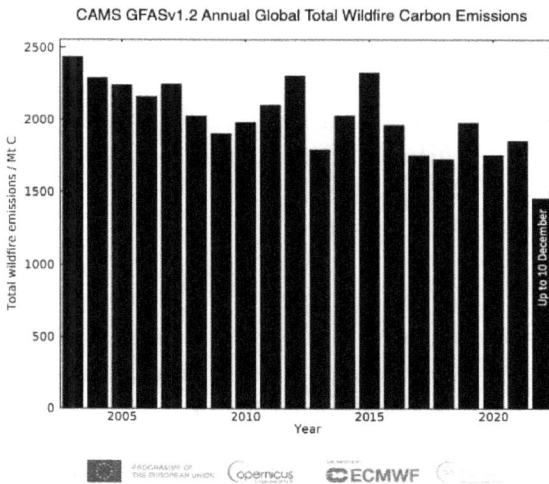

Figure 10-31 Global Total Wildfire Carbon Emissions [151]

127

These graphs do not suggest any causal link between the increasing concentrations of carbon dioxide in the atmosphere and wildfires.

Global Warming Process
Humans started burning fossil fuels around 1800, which added carbon dioxide to the atmosphere and heat to the environment. The effects of burning fossil fuels were identified around 1950, when the concentration of carbon dioxide in the atmosphere broke out of its relatively narrow range and increased substantially.

By all appearances, the Earth was sufficiently resilient to absorb the effects of burning fossil fuels in relatively small quantities for approximately 100 years with no apparent ill effects. However, the concentration of carbon dioxide in the atmosphere increased from approximately 280 to 420 ppm from around 1910 into the 2020s, with no end in sight. In addition, other greenhouse gases contribute approximately one-third of the greenhouse effect (see Figure 9-9).

Since around 1910, the higher than natural concentration of carbon dioxide in the atmosphere has trapped additional heat on Earth. In addition, burning fossil fuels produced heat that was dispersed into the environment. The greening of the Earth partially counteracted the accumulation of heat by providing a source of cooling, and the accumulation of carbon dioxide by increasing vegetation growth.

In 1977, the temperature of the Earth started to increase and rose approximately 1 degree Celsius by the early 2020s. The temperature of the Earth is projected to continue to rise as the burning of fossil fuels releases more carbon dioxide and heat into the environment. Melting polar ice and rising sea levels are largely consistent with an increase in the temperature of the Earth.

The Canary in a Coal Mine
Based upon the data presented, one can reasonably conclude that carbon dioxide is a greenhouse gas that, at elevated concentrations, causes the temperature of the Earth to rise, Arctic and Antarctic ice to melt, sea levels to rise, and other less dramatic and sometimes local changes. As such, the concentration of carbon dioxide in the atmosphere is like the canary in a coal mine that warns miners when the air in the mine is not safe. Stated differently, the concentration of

carbon dioxide in the atmosphere should be measured, monitored, and controlled, if possible.

However, carbon dioxide is a trace gas with a concentration of approximately 420 ppm that is increasing at a rate of approximately 2.31 ppm per year. The lowest OSHA occupational exposure limit for carbon dioxide is 5000 ppm time weighted average over 8 hours. [152] Exposure to levels above 1500 ppm can cause mild respiratory stimulation in some people, while exposure to levels below 1000 ppm typically has no symptoms but could cause drowsiness. [153]

If the Earth and its inhabitants can withstand a carbon dioxide concentration of 1000 ppm and its concentration continues to increase at the same rate, the concentration of carbon dioxide in the atmosphere would reach 1000 ppm in approximately 251 years {(1000-420)/2.31}. This rough calculation is not a prediction, but rather a straight-line calculation to estimate how much time humans might have to address the issue. Increasing carbon dioxide emissions in the future, the effects of indoor carbon dioxide accumulation, and accelerated concentration increases will reduce this time somewhat. Nonetheless, the relatively long period of time indicates that **the increasing concentration of carbon dioxide in the atmosphere is not an immediate existential crisis to human health**.

Putting this into perspective, humans burning fossil fuels for over 200 years increased the carbon dioxide concentration in the atmosphere from approximately 280 ppm to 420 ppm with a one-degree Celsius temperature rise. It would be reasonable to take a similar amount of time to mitigate and reverse these effects. Given the increasing rate of technical innovation that has occurred since around 1900, improving technology should enable mitigation of the effects of the increasing carbon dioxide concentration in the atmosphere. Ongoing research is focused on addressing some of these issues, including the development of processes that produce less carbon dioxide.

Mitigation of the effects of carbon dioxide accumulation in the atmosphere, such as the Earth's temperature rise, sea level rise, ocean acidity increase, snow-covered areas, polar ice, and ocean current changes, will not be immediate, but will follow decades after the carbon dioxide concentration in the atmosphere is slowed, stopped, and ultimately, reduced.

On a side note, the concentration of methane comprises approximately 15 percent of the AGGI as compared to approximately two-thirds for carbon dioxide (see Figure 9-12). Therefore, it would be reasonable to also consider mitigating human activities that emit methane into the atmosphere.

The Elephant in the Room

Global warming is often in the headlines, but global cooling may be the real existential threat. Every minute of every day of every year, the Sun generates an enormous amount of energy that is radiated in every direction. A small portion of that energy warms Earth and provides for our sustenance. Earth would be just a cold, barren rock floating around in space if not for the correct amount of energy provided by the Sun and retained on Earth.

The Sun constantly consumes fuel, so the amount of fuel remaining in the Sun will diminish, as will the amount of energy radiated by the Sun. Therefore, the amount of energy reaching Earth will diminish significantly in the long term. It is inevitable that this process will tend to make the Earth cooler.

A weak Sun can truly be an existential threat to the environment and human activity as we know it. What will happen when the energy from the Sun diminishes? Will humans and other living organisms adapt and be able to survive? What should we do now to prepare for this eventuality? Why is no one talking about this?

Importantly, the previous discussion did not mention that significant global cooling will occur a few billion years in the future. While some scientists do actively look for other inhabitable planets, it would not be prudent to worry about global cooling right now, or in the next few million years, for that matter. More importantly, this example illustrates how a real problem can be legitimately ignored when put into perspective.

The Conundrum

Carbon was stored underground long ago as coal, gas, and oil. Humans removed these fossil fuels from deep storage and burned them. The carbon dioxide produced was dispersed, increasing the concentration of carbon dioxide in the atmosphere. Less heat leaving Earth due to this increased concentration plus thermal processes releasing heat to the environment caused the Earth's temperature to increase.

The account in the previous paragraph may be based on science, but it is woefully lacking in perspective. Per Figure 6-1, significant use of coal started in the late 1800s and surpassed the use of biomass fuels around 1900, and nuclear power became commercially available in the 1950s. Electricity has facilitated many technological developments. Conversely, a less than ample supply of electricity stifles technological advances, including the development of nuclear energy.

If fossil fuels were never consumed, humans would be living with standards of living reminiscent of 1800, using rudimentary machines to make goods that would be limited in variety, quality, quantity, and sophistication. Plastics, synthetic rubber, artificial fibers, fertilizers, pesticides, and thousands of other materials and products made from fossil fuels would not exist. Medical devices and pharmaceuticals would be primitive, and not be that effective in curing disease, relieving pain, reducing suffering, and saving lives. Sadly, many parents might still be burying two of their five children before they reach their fifth birthday. In addition, we would not have the level of communication, transportation, water, food, sewage, electrical, health, educational, and entertainment infrastructures to which we have become accustomed to make our lives so much easier, more pleasant, and more productive. Stated differently, without fossil fuels, most of the global population would still be living in extreme poverty at Level 1, clinging to a 'life of death' that was the human condition for thousands of years.

On the other hand, if humans did not consume fossil fuel, the global population might be approximately one-fourth of its current size, the concentration of carbon dioxide in the atmosphere might be approximately 280 ppm, the temperature of the Earth might be approximately 1 degree Celsius cooler, the sea level might be approximately 10 cm lower, and the oceans might be a little less acidic.

If you are reading this book, you are probably living at Level 4 and have greatly benefited from the burning of fossil fuels. If you had a choice, would you choose your life now with electricity, radio, television, cars, airplanes, computers, cellphones, 3D printing, social media, and gene therapy, or a life of the 1800s when occasionally going hungry, using a latrine, gathering firewood, fixing things, and fetching water occupied a significant portion of daily life?

The high concentration of carbon dioxide in the atmosphere and its effects are unintended consequences of improving the standards of living for an increasing population.

Chapter 11: Climate Change and Climategate

Science is not about consensus, and consensus is not science. ---
Burt Rutan

Some years ago, I visited a small factory where the owner was an interesting sort. Just after 2:00 pm, he arranged for us to be with another person and disappeared. About 10 minutes later, he was spotted in a tee shirt, shorts, and tennis shoes heading for his daily ping pong match with an employee, returning about 45 minutes later to continue the meeting. It was his company, so he had the freedom to do as he wished. But I digress.

The owner was born under communist rule on the eastern side of the Iron Curtain. Early in life, he recognized that the news he received might be factual but was often woefully incomplete. Obtaining information from multiple sources to triangulate a complete news story became a game for him.

Raised in the United States, I did not really understand, so he provided an example. It seems that there was an international race. Factual enough. Russia came in second. Factual enough. Score a big one for Russia! After checking other news sources, he found out that the United States came in first. Factual enough. Score one for Russia anyway. After even more digging, he discovered that there were only two countries competing in the race. OOPS! Russia did not beat any country because it lost a head-to-head race.

The owner did his own research, because he knew better than to accept what he was told.

---xxx---

Discussions and technical papers about climate change and global warming are rampant with errors of omission, such as addressing only one aspect without mentioning other key aspects, providing incomplete information, and not presenting dissenting opinions, even when based on facts and data.

The starting point of many discussions is often an understanding that the science of climate change and global warming is already decided, so discussion often evolves into a circular argument concluding that the science of climate change and global warming is already decided, often without providing supporting data.

For example, statements that global warming affects vegetation are stated in a negative light and almost invariably omit the agricultural benefits that result from the greening of the Earth. Increased peace and prosperity that have been associated with a warmer climate are similarly not mentioned. **Rarely discussed are the thousands of products that are made from, or made with, fossil fuels that are considered essential to everyday life.**

Other discussions and technical papers contain errors. For example, details in the middle of a newspaper article entitled *Mama Mia! NYC rules crack down on coal-, wood-fired pizzerias — must cut carbon emissions up to 75%* describes a proposal to force restaurants with coal-fired and wood-burning ovens to hire an engineer or architect to assess the feasibility of installing emission control devices to achieve a 75% reduction in particulate emissions. 154

Carbon emissions and particulate emissions are not the same. Reducing carbon emissions by 75 percent suggests using a different fuel, which is not desirable to make pizza, or capturing carbon emissions, which is not practical. However, the mention of filters in the article and particulates in the statute implies that the intent is to address particulate emissions, not carbon emissions. Therefore, the title of this and many similar articles are incorrect, as were at least two reports on television, and one on the radio. That said, at least one article did point out the mischaracterization by the press and described the proposal in perspective. [155]

This chapter includes the evolution of global warming and climate change, and some of the international organizations ostensibly formed by scientists to study and address the issues. Also presented are dissenting examinations of the data that call into question whether the science of global warming is settled or not.

The Club of Rome

The Club of Rome is one of a "dazzling" array of elite organizations and think tanks tied to David Rockefeller (1915-2017), Chairman and Chief Executive Officer of Chase Manhattan Corporation, at the end of the 1960s and into the early 1970s. [156]

In April 1968, a group of thirty individuals from ten countries-scientists, educators, economists, humanists, industrialists, and national and international civil servants-gathered in the Accademia dei Lincei in Rome. They met... to discuss a subject of staggering scope - the present and future predicament of man.

Out of this meeting grew The Club of Rome, an informal organization that has been aptly described as an "invisible college." Its purposes are to foster understanding of the varied but interdependent components – economic, political, natural, and social – that make up the global system in which we all live; to bring that new understanding to the attention of policymakers and the public worldwide; and in this way to promote new policy initiatives and action. [157]

The Limits to Growth was a 1972 Club of Rome publication based on a model developed by a group of MIT scientists that examined the five basic factors that determine, and therefore ultimately limit, growth on this planet – population, agricultural production, natural resources, industrial production, and pollution. [158] Since the models can accommodate only a limited number of variables, the interactions studied are only partial. [159] Nevertheless, it was recognized that, with a simple world model, it is possible to examine the effect of a change in basic assumptions or to simulate the effect of a change in policy to see how such changes influence the behavior of the system over time. [160]

The Limits to Growth concluded that:

1. If the present growth trends in world population, industrialization, pollution, food production, and resource depletion continue unchanged, the limits to growth on this planet will be reached sometime within the next one hundred years. The most probable result will be a rather sudden and uncontrollable decline in both population and industrial capacity.

2. It is possible to alter these growth trends and establish a condition of ecological and economic stability that is sustainable far into the future. The state of global equilibrium could be designed so that the basic material needs of each person on earth are satisfied and each person has an equal opportunity to realize his individual human potential.

135

3. If the world's people decide to strive for this second outcome rather than the first, the sooner they begin working to attain it, the greater will be their chances of success. [161]

Preliminary views expressed in <u>The Limits to Growth</u> (1972) include interactions between demographic growth and economic growth that make it vital to consider alternatives to continued unrestricted material consumption, which compels mankind to seek a state of equilibrium that can be achieved only through a global strategy. Technological solutions alone cannot be expected to suffice, so the strategy for dealing with the two key issues of development and the environment must be conceived as a joint one. There is recognition that the complex world *problematique* is to a great extent, composed of elements that cannot be expressed in measurable terms. The authors are unanimously convinced that rapid, radical redressment of the present unbalanced and dangerously deteriorating world situation is the primary task facing humanity. This "supreme" effort is a challenge for our generation that cannot be passed on to the next. The effort must be resolutely undertaken without delay, and significant redirection must be achieved during this decade. This effort calls for joint endeavor by all peoples, whatever their culture, economic system, or level of development, but the major responsibility must rest with the more developed nations. We unequivocally support the contention that a brake imposed on world demographic and economic growth spirals must not lead to a freezing of the status quo of economic development in the world's nations. We affirm finally that any deliberate attempt to reach a rational and enduring state of equilibrium by planned measures, rather than by chance or catastrophe, must ultimately be founded on a basic change of values and goals at the individual, national, and global levels. [162]

Overall, the focus of <u>The Limits to Growth</u> in 1972 was to warn society that the Earth's resources are limited, and that immediate action needed to be taken to avoid undesirable, and perhaps catastrophic, outcomes. The increasing concentration of carbon dioxide in the atmosphere was mentioned, graphed from 1860, and projected to 2000. [163] Global warming was not mentioned because it had not yet started. Other topics discussed include population growth, pollution, environmental issues, and the limited availability of land, food, minerals, and metals.

The idea of population growth limits initially raised outrage from economists, industrialists, politicians, and Third World advocates. However, eventually, events demonstrated that the concept of global ecological constraints is not absurd. Resource and emission constraints have created many crises since 1972, exciting the media, attracting public attention, and arousing politicians. [164]

Regarding carbon dioxide in the atmosphere, The Limits to Growth states that:

> ...no upper bounds have been indicated for the exponential growth curves of pollutants ... because it is not known how much we can perturb the natural ecological balance of the earth without serious consequences. It is not known how much CO_2 or thermal pollution can be released without causing irreversible changes in the earth's climate... [165]

In 1974, The Club of Rome called for the development of a master plan to allocate resources in a new global order, further stating that nations would lose at least some of their independence.

> Now is the time to draw up a master plan for organic sustainable growth and world development based on global allocation of all finite resources and a new global economic system. [166]

> Increasing interdependence between states and regions must then translate as a decrease in independence. Nations cannot be interdependent without each of them giving up some of, or at least acknowledging limits to, its own independence. [167]

Interestingly, the 30-year update in 2004 states that The Club of Rome research does not try to predict the future but is rather an effort to identify different possible futures by sketching alternative scenarios for humanity. [168] This appears somewhat disingenuous, because the predicted population growth formed the underlying foundation of The Club of Rome's original thesis.

United Nations Conference on the Human Environment (1972)

With respect to carbon dioxide, Recommendation 57 of the wide-ranging Action Plan for the Human Environment from this meeting in Stockholm, also called Earth Summit I, recommends:

> ...that the Secretary-General take steps to ensure proper collection, measurement and analysis of data relating to the environmental effects of energy use and production within appropriate monitoring systems.
>
> (a) The design and operation of such networks should include, in particular, monitoring the environmental levels resulting from emission of carbon dioxide, sulphur dioxide, oxidants, nitrogen oxides (NO_x), heat and particulates, as well as those from releases of oil and radioactivity;
>
> (b) In each case the objective is to learn more about the relationships between such levels and the effects on weather, human health, plant and animal life, and amenity values. [169]

United Nations Conference on Environment and Development (1992)

The primary objective of the Rio 'Earth Summit' was to produce a broad agenda and a new blueprint for international action on environmental and development issues that would help guide international cooperation and development policy in the twenty-first century to focus on the impact of human socio-economic activities on the environment. This following excerpt from the opening statement given by Maurice Strong (1929-2015), Secretary General of the Rio Summit and friend of David Rockefeller, includes both Rosling's *fear instinct* and the Malthusian worldview that the population was becoming too large for resources to support.

> We must honour, and the Earth must support, all its children. But, overall, this growth cannot continue. Population must be stabilized, and rapidly. If we do not do it, nature will, and much more brutally.
>
> We have been the most successful species ever; we are now a species out of control. Our very success is leading us

to a dangerous future. The concentration of population growth in developing countries and economic growth in the industrialized countries has deepened, creating imbalances which are unsustainable, either in environmental or economic terms. [170]

Debunked: Strong later said, *Isn't the only hope for the planet that the industrialized civilizations collapse? Isn't it our responsibility to bring that about?* [171] This appears to be a revolutionary call to destroy our civilization. After considerable research, the origin of this statement turned out to be in the last column on the last page of a 10-page magazine article published in 1990 about his life, where Strong talked about a novel that he wanted to write. Although quoted by others, this statement should be disregarded because it refers to fiction and not fact.

What is written, what is spoken, and what is implied must all be questioned. Statements can be factual, but investigation is required to evaluate the context.

International Treaties

The Intergovernmental Panel on Climate Change (IPPC) was established in 1988 to provide policymakers with regular scientific assessments on climate change, its implications, and potential future risks, as well as to put forward adaptation and mitigation options. [172] The IPPC has been issuing periodic assessment reports since 1990.

The 1992 United Nations Framework Convention on Climate Change (UNFCCC) was established to support the global response to the threat of climate change and prevent human activity from affecting the Earth's climate, in part by stabilizing greenhouse gas emissions. Agenda 21 is a related action plan to address sustainable development.

The Kyoto Protocol, adopted in 1997, extended the UNFCCC and contained targets for specific countries to limit and reduce their greenhouse gas emissions.

The Paris Agreement, also known as the Paris Accord or Paris Climate Accord, was adopted in 2015 as an international treaty with the goal of limiting the increase in the global average temperature to well below 2 °C above pre-industrial levels by reaching net zero carbon dioxide emissions by around 2050. Reaching the preferred 1.5 °C rise entails cutting emissions by 50 percent by 2030. The Paris Agreement contains provisions to help countries adapt. The agreement has been criticized by environmentalists and analysts because no emissions targets were set.

Climategate

Previous analysis based on data from credible sources shows that global warming exists. However, flawed data and inappropriate manipulation of that data are legitimate reasons to investigate the techniques used to obtain the raw data, how the raw data was manipulated, and, if flawed, to present corrected information and an updated analysis.

S. Fred Singer (1924-2020), a physicist and professor of environmental science, wrote books and many contemporaneous publications that provide what appears to be credible evidence of flaws in the data and its analysis. Singer suggests that there were several corruptions in the peer-review process, [173] but was particularly critical of Figure 11-2, the "hockey stick" graph in the 2001 IPCC Third Assessment Report, as compared to Figure 11-1 in the 1990 IPCC Scientific Assessment. Figure 11-2 contains the same temperature data as Figure 10-5, which was used to show that significant global warming is occurring.

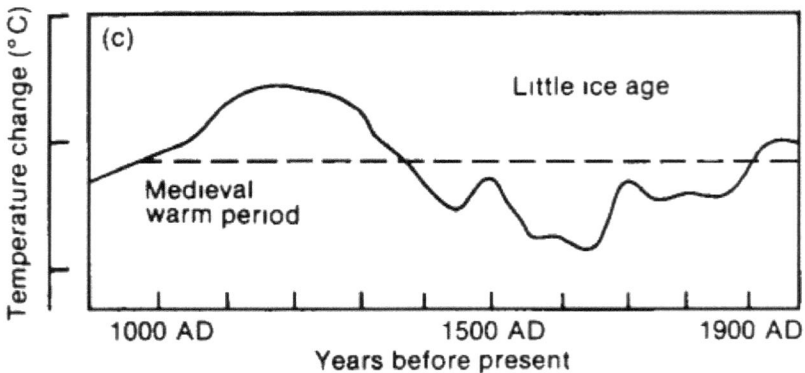

Figure 11-1 IPCC Temperature Data (1990) [174]

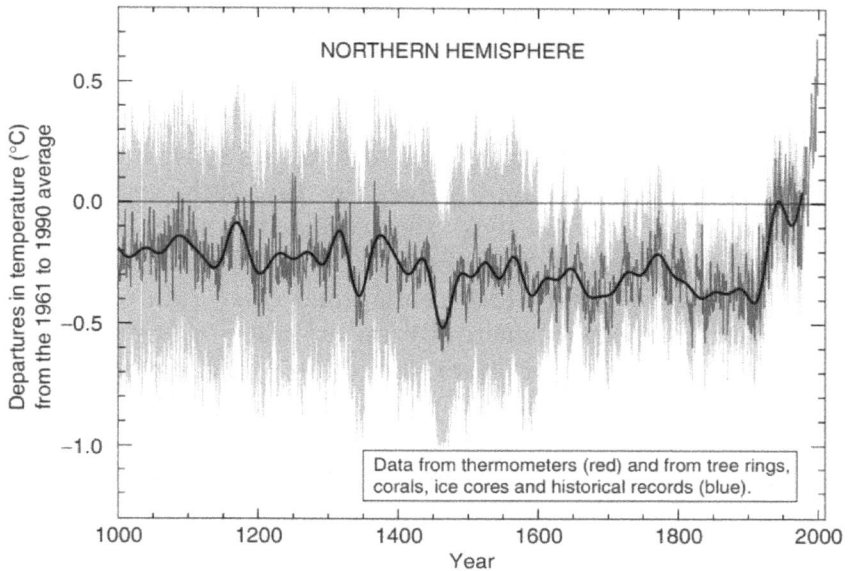

Figure 11-2 IPCC Temperature Data (2001) [175]

The "hockey stick" graph primarily used tree rings as a basis to assess past temperature changes, in conjunction with more recent surface instrument records. This effectively replaced the temperatures during the Medieval Warm Period (around 1100) and the Little Ice Age (around 1400 to 1850) with relatively stable temperatures. Then, twentieth century temperatures "seem to rocket upward out of control" implying that the warming of the twentieth century was unprecedented or unnatural, but this is wrong. [176] In response, the IPCC suggested that the Medieval Warm Period was not a globally uniform event, but rather a localized event. [177]

The conclusion that the "hockey stick" graph is wrong is supported in the following abstract from a research investigation. (**Emphasis added**)

The data set of proxies of past climate used in Mann, Bradley and Hughes (1998, "MBH98" hereafter) for the estimation of temperatures from 1400 to 1980 contains collation errors, unjustifiable truncation or extrapolation of source data, obsolete data, geographical location errors, incorrect calculation of principal components and other

141

quality control defects. We detail these errors and defects. We then apply MBH98 methodology to the construction of a Northern Hemisphere average temperature index for the 1400–1980 period, using corrected and updated source data. **The major finding is that the values in the early 15th century exceed any values in the 20th century.** The particular "hockey stick" shape derived in the MBH98 proxy construction – a temperature index that decreases slightly between the early 15th century and early 20th century and then increases dramatically up to 1980 — is primarily an **artefact of poor data handling, obsolete data, and incorrect calculation of principal components**. [178]

Singer reported that the researchers demonstrated that even when data without trends is entered into the formula, hockey-stick-shaped patterns result. [179] Of circumstantial interest, the quality and manner in which the data was received indicated that no one else had requested the data and the absence of a true peer review. In addition, Singer questions the accuracy of surface temperature measurements. [180]

In November 2009, a whistleblower leaked thousands of emails involving high-ranking scientists in the IPCC that exposed the "manhandling of fundamental data" to "hide the decline" in air temperature post-1980. A second batch of emails was released in 2011, and many emails "clearly confirm that top IPCC scientists consciously misrepresented and actively withheld information…then attempted to prevent discovery." [181] Singer quoted another individual as saying that the IPCC is an "organized conspiracy dedicated to tricking the world into believing that global warming is a crisis that requires a drastic response." [182] Of circumstantial interest, the developer of the "hockey stick" graph sued another researcher for defamation but refused to produce the original data during pre-trial discovery and was ordered to pay the researcher's legal fees. [183]

In other words, the **data that formed the fundamental basis of global warming is being credibly questioned**. Numerous other disagreements with statements made about global warming and climate change have been published by the Nongovernmental International Panel on Climate Change (NIPCC).

The Limits to Growth Updates

The 30-year update of The Club of Rome The Limits to Growth recognized that there had been many positive developments since 1972, including increased awareness of environmental matters, the decline in population growth in response to increased income, the reduction of pollution from smokestacks and outflow pipes of factories in the rich world, a push for ever higher eco-efficiency and sustainability, and growth rates of food, energy, and industrial production that far exceeded population growth in some regions. Negative developments included increased greenhouse gas emissions, and 54 nations, representing 12 percent of the global population, experienced gross domestic product (GDP) decreases in the 1990s. The report states that the "global challenge can be simply stated: To reach sustainability, humanity must increase consumption levels of the world's poor while at the same time reducing humanity's total ecological footprint … (that) will take decades to achieve even under the best of circumstances." [184]

> The bad news is that many crucial sources are emptying and degrading, and many sinks are filling up or overflowing. *The throughput flows presently generated by the human economy cannot be maintained at their current rates very much longer.* The good news is that the current *high rates of throughput are not necessary to support a decent standard of living for all the world's people.* The ecological footprint could be reduced by lowering population, altering consumption norms, or implementing more resource-efficient technologies. [185]

Further stated is that, if the industrial world begins to act upon two definitions of *enough*, one having to do with material consumption, the other with family size, the world population can achieve a level of well-being roughly equivalent to the lower-income nations of present-day Europe, [186] which is comprised of the Eastern European countries located between and including Poland and Greece, whose median income is approximately 30 to 40 percent that of the United States. [187] There is also the suggestion of transitioning to a sustainable system whereby poverty,

143

unemployment, and unmet material needs require new thinking, [188] with visioning, networking, truth-telling, learning, and loving tools. [189]

The 50-year update contains 21 essays written by The Club of Rome members, university professors, and others to examine what has occurred since 1972 and provide suggestions about the future. This 2022 update contains little technical information but does contain an entire chapter describing a shift from mainstream capitalism to *solidarity capitalism.*

> Given the planetary emergency we find ourselves in ... we must strike a new balance, considering not just those who demand and can afford access to resources, but also those who have needs and good cause to make demands but are denied their seat at the table. "Solidarity Capitalism", in this context, speaks to a new understanding and approach to managing our economic and financial systems that becomes essential if humanity is to collectively address the challenge of climate change ... We will explore what is required to enable the implementation of such a model by shifting from individualistic capitalist ideology to one of solidarity, wherein the common good is safeguarded in tandem with the pursuit of self-interest. [190]
>
> Promoting Solidarity Capitalism is not about introducing the world to an alternative terminology for capitalism, but more about embedding the ideas of solidarity and social responsibility into the ideals and cultures of how businesses and economic systems are managed. ... [it] is about changing mindsets, re-engineering the system, building an internal framework that draws from a multiverse of perspectives and sets the minimum benchmark for innovation to be in the service of Life at large. [191]
>
> This creates new possibilities of economic access, equity, social justice, facilitating connection, and a sense of belonging for all – solidarity. [192]
>
> Solidarity Capitalism as an idea, or academic concept, would encourage autonomous thinking by aligning the sense of community and social responsibility with the capitalist model. [193] Solidarity Capitalism as an ethical concept, as

144

well as an economic model, ensures that we accompany the switch from a shareholder-centric to a stakeholder-focused model with ethical principles that include fairness and equity. [194]

Solidarity Capitalism is not just about the terminology, but rather about seizing the opportunity to merge the old with the new by allowing a multiverse of viewpoints to truly cross-pollinate. It is about achieving the interests of business while keeping the interests of the community and society in mind. [195]

To remodel modern capitalism as a system that serves humanity and the planet... [196]

In 2023, The Club of Rome website describes itself as a "platform of diverse thought leaders who identify holistic solutions to complex global issues and promote policy initiatives and action to enable humanity to emerge from multiple planetary emergencies." [197] **Since its formation, The Club of Rome, its publications, and its members, have had a considerable influence on global warming conferences and treaties.**

However, not everyone was enthralled with The Limits to Growth (emphasis not added).

In 1971, The Club of Rome published a deeply flawed report, **Limits to Growth**, which predicted an end to civilization as we know it because of rapid population growth, combined with fixed resources such as oil. The report concluded that without substantial changes in resource consumption, **"the most probable result will be a rather sudden and uncontrollable decline in both population and industrial capacity."** It was based on bogus computer simulations by a group of MIT computer scientists. [198]

In fairness, the characterization of "deeply flawed", while having the benefit of an additional 50 years of data, is perhaps a bit harsh. After all, Figure 4-1 shows that population growth was exponential in 1972. As it turned out, the 1972 projections of the population and carbon dioxide concentration in 2000 were reasonably accurate. However, the projection in Chapter 4 for the global population to grow rapidly until approximately 2060, slow dramatically, and then stabilize suggests that the Rosling *straight-line instinct* may have played a role in the 1972 projection and that the declining fertility rate, led by Brazil and the United States in the 1960s shown in Figure 4-7, was not considered.

Global Temperature Rise

Satellites have been used to measure tropical mid-tropospheric temperature variations since 1979. Figure 11-3 shows that there have been significant differences between the mathematical models used to forecast these variations and satellite measurements.

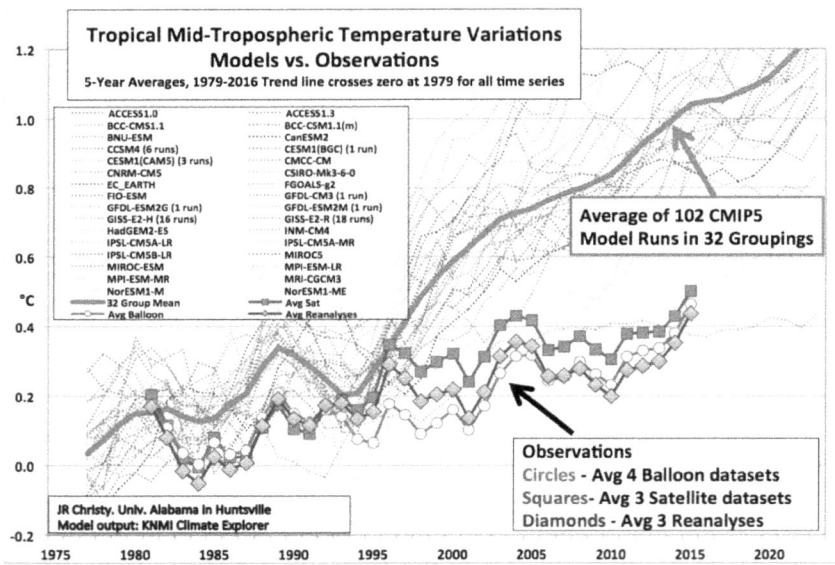

Figure 11-3 Tropical Mid-Tropospheric Temperature Variations [199]

Between about 1979 and 2015, Figure 11-3 and Figure 11-4 show net warming of approximately 0.3 °C and 0.6 °C, respectively. However, these charts represent temperatures at different locations. Figure 11-3 is based on tropospheric temperature measurements, whereas Figure 11-4 is based on surface temperatures. Tropical mid-tropospheric temperature variations are "particularly important because they capture the atmospheric region that is anticipated to warm rapidly and unambiguously if greenhouse theory is well understood. As such, if the impact of extra greenhouse gases (GHGs) is to be detected, it should be detected here." [200] This brings

146

into question whether Figure 10-5 should be used to analyze global warming, especially after 1979.

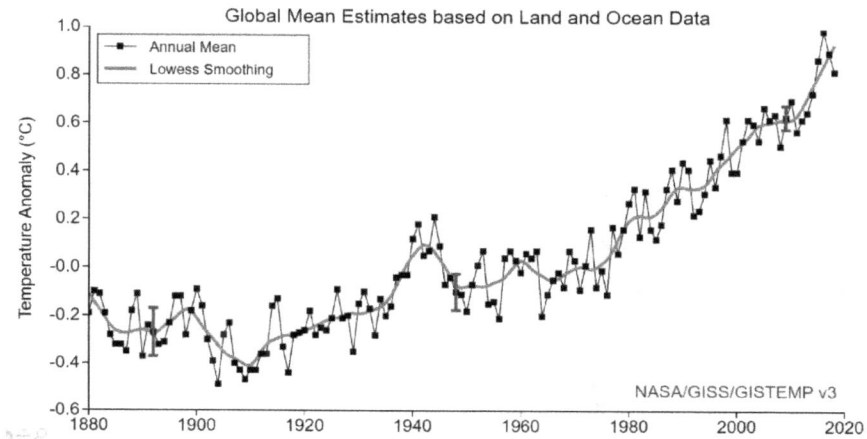

Figure 11-4 Global Mean Surface Temperature Variations [201]

In addition, the disappearance of the Medieval Warming Period and the Little Ice Age, which collectively represent over half of Figure 10-5, leads one to question the accuracy of the remainder of Figure 10-5 and any conclusions drawn from it.

Further, Singer references Figure 11-5, which in addition to showing the Medieval Warming Period and the Little Ice Age, also shows that temperature variations since 1800 are not extreme and that there is nothing unnatural about the current temperature of the Earth. [202] It also shows global cooling after around 1940, which was concerning at the time.

Sunspots are bursts of energy emitted by the sun. Sunspot activity peaks approximately every 11 years. Times of maximum sunspot activity are associated with a very slight increase in the energy output from the sun [203] that tends to increase the global temperature. Sunspot maximums gradually fell to less than half of their 1957 peak by 2014. [204] This suggests that the global temperature rise might have been higher since 1957, if sunspot activity had not fallen.

Earthquakes and volcanic eruptions block the sun and decrease the amount of energy received by the Earth, which tends to decrease the global temperature.

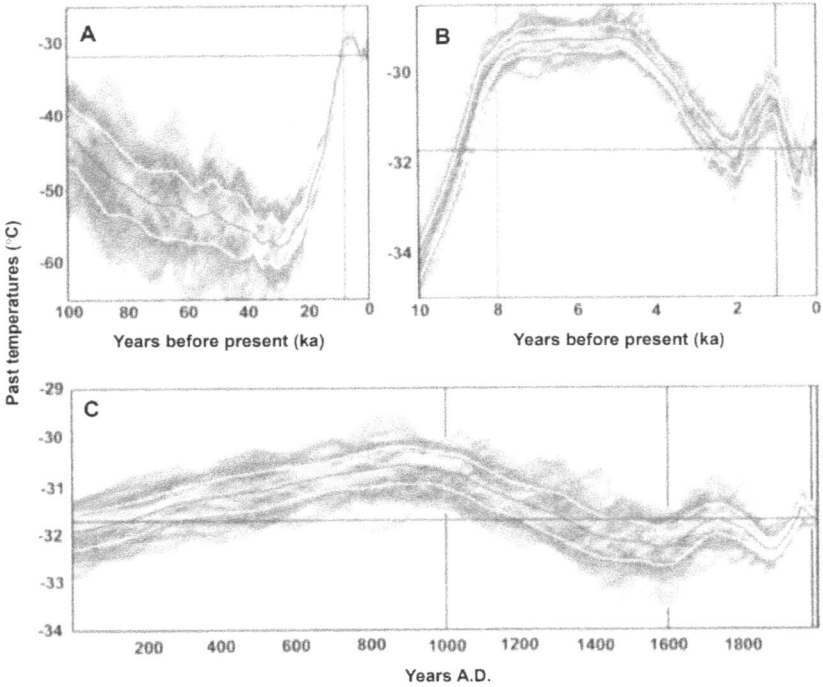

Figure 11-5 Temperature Data from the Greenland Ice Core Project
(GRIP) [205] [C]

Representation Concentration Pathway (RCP)

The difference between the predicted and measured temperatures in Figure 11-3 shows that 31 of the 32 models significantly overstated the measured temperature rise. This does not bode well for models and projections.

Figure 11-6 was presented at the 2014 United Nations Intergovernmental Panel on Climate Change (IPCC) to show the projected carbon dioxide concentrations in the atmosphere under different Representation Concentration Pathways (RCP) scenarios.

[C] Graph A shows temperature data 100,000 years ago to present. Graph B shows details data 10,000 years ago to present, which is in the upper right corner of Graph A. Graph C details data 1000 years to present, which is on the right side of Graph B. Stated differently, the graphs show increasing detail in increasingly recent times.

148

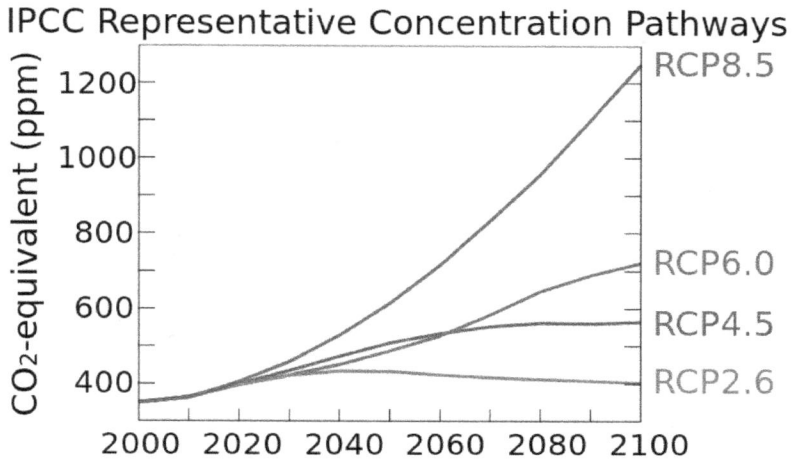

Figure 11-6 IPCC Representative Concentration Pathways (2014) [206]

Information from the IPCC website is instructive. (**Emphasis added**)

> Four RCPs were selected and defined by their total radiative forcing (cumulative measure of human emissions of GHGs from all sources expressed in Watts per square meter) pathway and level by 2100. The RCPs were chosen to represent a broad range of climate outcomes, based on a literature review, and *are neither forecasts nor policy recommendations*.

> While each single RCP is based on an internally consistent set of socioeconomic assumptions, the four RCPs together cannot be treated as a set with consistent internal socioeconomic logic. For example, *RCP8.5 cannot be used as a no-climate-policy socioeconomic reference scenario* for the other RCPs because RCP8.5's socioeconomic, technology, and biophysical assumptions differ from those of the other RCPs.

> Each RCP could result from different combinations of economic, technological, demographic, policy, and institutional futures. For example, the second-to-lowest RCP could be considered as a moderate mitigation scenario. However, it is also consistent with a baseline scenario that assumes a global development that focuses on technological improvements and a shift to service industries but does not

aim to reduce greenhouse gas emissions as a goal in itself (similar to the B1 scenario of the SRES scenarios). [207]

The report made significant errors, [208] and flawed papers and research based on RCP 8.5 are now the norm, which has accelerated efforts to eliminate natural gas and nuclear for intermittent electricity from wind turbines and solar panels. [209] For example, a graph remarkably similar to Figure 11-6 appears in a research paper predicting that the number of home runs in baseball will increase because of global warming. [210]

Stated differently, despite models significantly overstating the temperature rise, the most extreme scenario in a flawed study was and is still being used to draw attention, show urgency, and form the basis of a global policy that is replacing reliable electricity generation from fossil fuels, with unreliable electricity generation from solar and wind sources.

Carbon Dioxide Measurements

Zbigniew Jaworowski studied the ice core and other methodologies used to obtain historical carbon dioxide measurements and questioned the ice core carbon dioxide measurements that form the basis of the IPCC reports. [211] Figure 11-7 shows how the original ice core measurements were shifted 83 years to provide continuity with the actual atmospheric measurements at Mauna Loa, Hawaii, in 1958. The IPCC used a curve similar to the shifted curve on the right.

FIGURE 2(a) and (b)
Mother of All CO$_2$ Hockey Curves

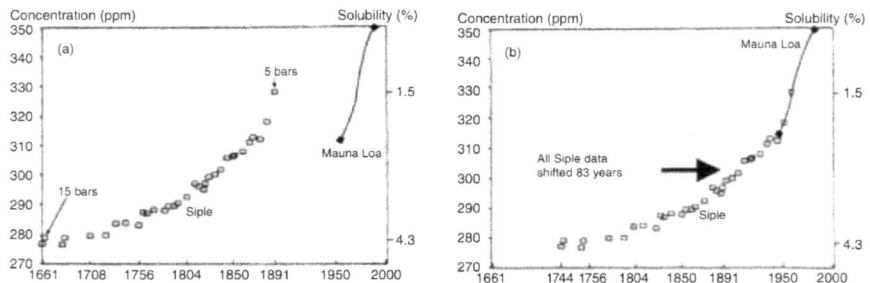

Source: Adapted from Friedli et al. 1986 and Neftel et al. 1985.

Concentration of CO$_2$ in air bubbles from the pre-industrial ice from Siple, Antarctica (open squares), and in the 1958-1986 atmosphere at Mauna Loa, Hawaii (solid line). In (a), the original Siple data are given without assuming an 83-year-younger age of air than the age of the enclosing ice. In (b), the same data are shown after an arbitrary correction of the age of air.

Figure 11-7 Unshifted and Shifted Carbon Dioxide Concentrations [212]

Figure 11-8 shows a reconstruction of atmospheric carbon dioxide concentration based on direct chemical measurements, which is remarkably different from ice core measurements and Figure 11-7.

FIGURE 3
First Reconstruction of Trends in CO₂ Atmospheric Concentration Based on Actual Measurement

CO2 -1812 - 2004 Northern Hemisphere , Chemical Measurement

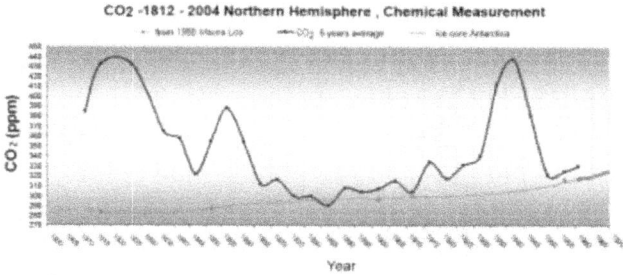

This first reconstruction of trends in CO₂ concentration in the Northern Hemisphere is based on more than 90,000 direct chemical measurements in the atmosphere at 43 stations, between 1812 and 2004. The lower line are the values from Antarctic ice core artifacts. The diamonds on the lower line (after 1958) are infrared CO₂ measurements in air from Mauna Loa, Hawaii.

Source: Adapted from Beck 2007.

Figure 11-8 Carbon Dioxide Trends Based on Measurements [213]

The data presented in Figure 11-8 is in stark contrast to Figure 9-5, which shows that the carbon dioxide concentration was relatively stable during the last 2000 years. If the reconstruction based on measurements in Figure 11-8 is accurate, then the carbon dioxide concentration prior to around 1800 varied between approximately 290 and 450 ppm. This implies that the carbon dioxide concentration of 420 ppm in 2023 is within a normal range, which **puts into question whether the carbon dioxide concentration is increasing due to human activity**.

Temperature and Solar Cycle Length

Zbigniew Jaworowski suggests that Figure 11-9 shows that the driving force of temperature fluctuations was caused by solar activity rather than by CO_2 changes, which lag the temperature changes and appear not to be the cause of the temperature variations. [214]

This **puts into question whether variations in the Earth's temperature are related to the carbon dioxide concentration in the atmosphere**.

FIGURE 6
Average Northern Hemisphere Temperature

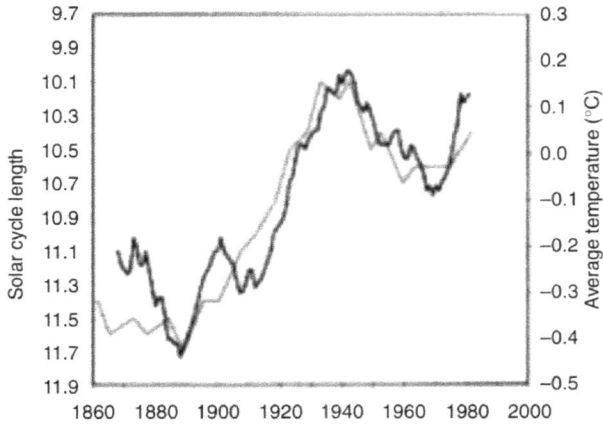

Source: After Friis-Christensen and Lassen 1991.

Figure 11-9 Average Northern Hemisphere Temperature and Solar Cycle Length [215]

Temperatures Rising on Other Planets

Some people claim that the temperature of other planets has been rising in conjunction with the temperature rise on Earth. If true, this would imply that the temperature rise on Earth is occurring naturally and not necessarily because of human activity. Reliable data to determine the validity of these claims could not be found.

Does Global Warming Exist?

The previous chapter presented data and concluded that increasing carbon dioxide in the atmosphere caused the temperature of the Earth to increase. This chapter presents additional information, data, and analysis that undermined much of the foundation on which the conclusions in Chapter 10 were based. This begs the question of which analysis is correct.

There is virtually universal agreement that humans released a large quantity of carbon dioxide into the atmosphere and that it was only partially absorbed by the Earth. The concentration of carbon dioxide in the atmosphere is increasing, as shown by the measurements in Figure 9-6. These increases pose significant concerns because, barring the appearance of an unknown self-regulating mechanism or the adoption of an unknown new energy

152

source with low emissions, the concentration of carbon dioxide in the atmosphere is on a path to eventually reaching levels that can adversely affect human health.

Pragmatically, it does not matter if global warming exists or, if it does, whether it is caused by human activity, because the underlying cause of both adverse health effects and global warming (if it exists) is the same, that is, the increasing concentration of carbon dioxide in the atmosphere. Addressing the increasing concentration will mitigate both issues. In analogy, if a doctor treats a wound on your hand, it does not matter whether the knife that caused the wound was clean or dirty, because the doctor will thoroughly clean the wound, and apply antibiotics in either case. The same is true for Earth, where the common remedy is to reduce carbon dioxide emissions.

What matters is determining which actions should be taken to control the concentration of carbon dioxide in the atmosphere and, importantly, how quickly they should be implemented. Of particular importance is acting in a way that causes minimal or no disruption to the global population.

Just as important is the determination of how quickly actions should be implemented, given the technology available and in development. Acting too slowly pushes the Earth closer to its limit, whereas acting too quickly can cause catastrophic negative consequences for the global population. Stated differently, there are tradeoffs between which actions are implemented, how fast they are implemented, and the speed with which the concentration of carbon dioxide in the atmosphere moves toward the desired concentration to sustain life on Earth.

Chapter 12: Approaches to Reduce Carbon Emissions

Think outside the box. --- Popular Metaphor

There was a rude awakening awaiting me on my first day of graduate school. Having spent four years studying electrical engineering, my first graduate course in systems engineering was taught by none other than the author of the textbook used in an advanced systems course at my undergraduate university.

Arriving early for the first day of class allowed me time to have a brief conversation with another student that quickly made my need to expand my horizons clear. My undergraduate degree was in electrical engineering, with additional course work in systems engineering. I was under the impression that studying systems in the electrical engineering department at the graduate level would require at least an electrical engineering background. However, the other student had no electrical engineering or systems background. He was an economics major. How did he even get into this class? How could he understand the material?

What I learned was that the economy is a system that has inputs and outputs, just like an electrical system. Some inputs related to the economy include interest rates, capital investment, and government spending. Changing these economic inputs affects the outputs of the economy, some of which are the gross domestic product, employment, and exports. Complex mathematical models relating the inputs and outputs of the economy have been developed and incrementally improved to make the models better match the actual economy. Once operational, running the model with different inputs can provide policymakers with insights into the probable effects of making changes. With this information in hand, policymakers can better determine how to adjust the real inputs to achieve the desired outputs in the real economy.

---xxx---

An actual model of the economy is extremely complex. However, the takeaway from my discussion with the economics major was that the economy is a system that can be conceptualized as a rectangular box with multiple inputs on one side and multiple outputs on the other. Importantly, most of the inputs have a negligible effect on the outputs, whereas a small number of inputs can have profound effects.

For example, how much money one person spends each month has virtually no effect on the overall economy, whereas changing interest rates can have profound effects. Interestingly, the amount of money spent each month by everyone in aggregate can also have profound effects. Understanding the process often allows the many inputs and outputs that have little or no effect to be removed from the analysis without sacrificing the quality of the results.

Similarly, considering global warming as a system can help provide a deeper understanding of the process. The inputs and outputs of the global warming system and the economic system are different, but the overall concept is the same. As with the economic system, considering all the many inputs and outputs of global warming would be both a monumental and futile task. However, the many inputs and outputs that do not significantly affect global warming need not be considered, thus greatly simplifying the analysis. Conversely, the few inputs that can strongly affect global warming should be fully investigated.

The most significant issue at hand is that the concentration of carbon dioxide in the atmosphere has increased dramatically since around 1950. Whether this caused the Earth to warm and adversely affect the climate is not important at this time because the accumulation of carbon dioxide in the atmosphere will adversely affect human health in around a century or sooner, whereas the other indirect effects, such as storm intensity, bear populations, and ocean temperature generally take much longer or are of lesser importance.

What does matter is reducing carbon dioxide emissions and perhaps removing carbon dioxide from the atmosphere until the Earth can be stabilized within a desired range that does not adversely affect human health or the environment. Measuring the concentration of carbon dioxide in the atmosphere provides a window into the global warming process and can be used to gauge progress.

Figure 8-1 shows that petroleum, natural gas, and coal fossil fuels represent 79 percent (36+32+11) of total energy consumption. Fossil fuels are also the largest source of carbon dioxide emissions.

It was previously noted in Figure 9-9 that the carbon dioxide concentration in the atmosphere started to increase around 1910, suggesting that the Earth could handle approximately 3 billion tons of carbon dioxide that were then emitted annually. Figure 9-9 also shows that the carbon dioxide emissions in the mid-2020s will be approximately 37 billion tons per year. **Returning to pre-1800 emission levels would ultimately entail reducing emissions by over 90 percent!**

One approach to reducing the concentration of carbon dioxide in the atmosphere is to reduce fossil fuel consumption by reducing end-use. There are various means by which this can be implemented.

Discontinue Burning Fossil Fuels

Fossil fuels are the major source of carbon dioxide emissions by humans, so one potential strategy is to immediately ban the burning of all fossil fuels, including coal, oil, and natural gas. There are issues with this approach at many levels.

Per Figure 8-1, mothballing power plants that burn coal, natural gas, and oil would reduce electrical power generation capacity by approximately 50 percent when the sun is shining and the wind is blowing, and by approximately 70 percent when not. Most people would need to find an alternative fuel to heat their homes or go cold. Air conditioning would become a thing of the past. Farm vehicles would have little fertilizer and no fuel, so agricultural production would suffer, and people might starve. Water and wastewater companies would operate at limited capacity with reduced hours, so people would not have to fetch water every day, assuming that there is water nearby. Industry would operate on a severely reduced schedule or shut down for lack of energy and raw materials. India experienced some of these eventualities in the mid-2020s.[216]

There would be no fuel for transportation, so the economy would come to a standstill. The list goes on and on, and daily life would quickly devolve back towards a Level 1 existence. On a side note, **fossil fuels provide not only energy, but also form the foundation for or enable the production of thousands of everyday products that would not otherwise exist, including asphalt, tar, plastics, metals, fertilizers, cosmetics, medicines, and almost everything else commonly used in daily life.**

Discontinuing the burning of fossil fuels is not a viable strategy.

Increase the Cost of Fossil Fuels

It is often said that consumption of a particular product increases when it becomes less expensive and more convenient to use. However, there are many instances where this is not the case.

Mitigating the adverse effects of energy consumption often falls on those who do not use that energy. For example, homeowners burn natural gas to heat their homes and receive benefits from burning fuel that results in its products of combustion being emitted into the atmosphere. However, other people who did not receive any benefit are effectively left to deal with the consequences, such as the accumulation of carbon dioxide in the atmosphere and subsequent global warming.

A *carbon tax* has been proposed to reduce carbon emissions by utilizing market forces by levying a tax on the purchase of fossil fuels to ensure that the consumer who benefits pays for both the fuel and its cleanup. Revenues raised can be used to fund research and reduce other taxes. This structure is like that of gasoline taxes in the United States, which are collected and used for highway repair, maintenance, and infrastructure projects. On a side note, gasoline taxes in the United States will need to be restructured as tax revenues decrease due to the adoption of electric vehicles.

However, the carbon tax is intended to disincentivize consumption, so it needs to be high, sometimes excessively high, which can adversely affect economic activity and the population. Carbon taxes, in and of themselves, do not reduce emissions but rather rely on market forces to reduce consumption.

Figure 12-1 and Figure 12-2 show that gasoline supplied dropped by approximately 10 percent after COVID-19 restrictions were lifted in 2021 as compared to 2019, even though the price of gasoline was 50 to 100 percent more expensive during this time. To gain additional perspective, keep in mind that Figure 12-2 shows average retail gasoline prices peaking at approximately USD 5.00 per gallon. Prices in California reached over USD 7.00 per gallon.

Thousand Barrels per Day

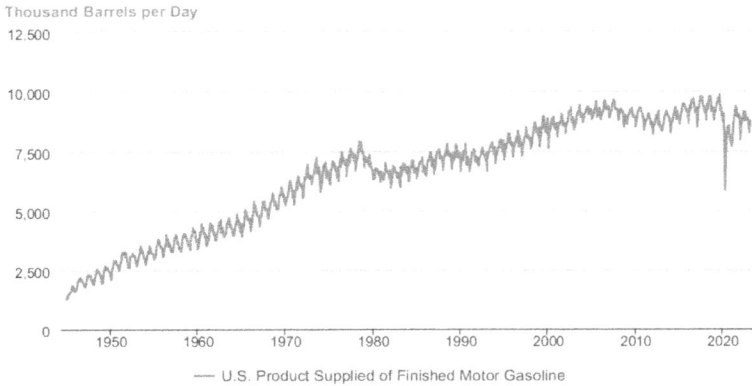

Figure 12-1 U.S. Gasoline Supplied by Month [217]

U.S. All Grades All Formulations Retail Gasoline Prices

Dollars per Gallon

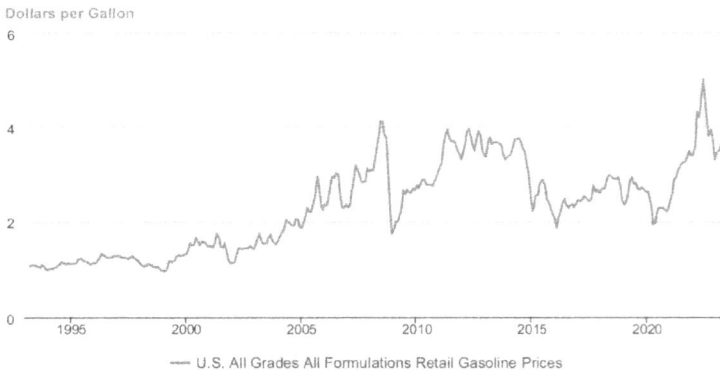

Figure 12-2 Retail Gasoline Prices [218]

High gasoline prices might pose an inconvenience for people living at Level 4, but life becomes increasingly difficult for those living at the lower levels. High prices significantly affect their standard of living because they need to buy goods and services that also become more expensive when suppliers pass the higher taxes for their fuel-related purchases along to consumers. Therefore, the carbon tax burden would affect everyone, but especially energy-intensive industries and people living below Level 4.

High prices might cause people to consume less gasoline, but most people will still use their vehicles to take their children to activities, go to the store, and go to work. Economic activity will be adversely affected when necessity curtails eating in restaurants, pleasure drives, family trips, entertainment, and vacations. Based on the data that high prices resulted in an approximately 10 percent gasoline reduction, a more meaningful reduction of, for example, 50 percent will require exorbitantly high prices, in which case both people and economic activity will suffer dramatically and perhaps catastrophically.

COVID-19 restrictions enabled some people to work from home. However, if the price of gasoline increases excessively, people who are required to work onsite might have to pay the high prices or find another job closer to home that might come with a lower income. They could also choose to relocate close to their place of work, as was done in the 1800s. Another alternative is for them to reduce their standard of living by a level or two.

Therefore, **raising fossil fuel prices is not a viable strategy** because it will cause the population to suffer and not appreciably reduce the concentration of carbon dioxide in the atmosphere.

Raising the price of other fuels can also be an exercise in futility. For example, natural gas is easier to use and much more convenient than gathering firewood for cooking, as was done for centuries. How much less food would the average person cook if the cost of natural gas was suddenly increased for cooking? Not much less, if any. How low would you set your heat to keep from being cold in the winter? A little lower, maybe. People may be able to save a small amount of energy, but in the end, they will use what they must.

Energy demand in the residential and commercial sectors is ultimately set by end-user demand and not by the energy provider. Similarly, user demand sets energy consumption in the industrial sector. However, energy prices might help incentivize an increase or decrease in production that will increase or decrease the demand for energy, respectively.

In short, the demand for energy is largely determined by end-users, and not energy providers. Raising the cost of fossil fuels may reduce energy demand somewhat, but it would put a heavy burden on the population. Higher transportation and residential sector energy costs would ripple through the economy and cause inflation, because virtually all products have fossil fuel components in their production and transport to market.

As is often the case during hardship, the most acute effects of high fossil fuel prices will be felt by people living at Levels 1, 2, and 3. People living at Level 4 might grumble about high prices, but the remainder of the population, who can least afford the added expenses, might be devastated. In practice, fossil fuel energy consumption will not drop much as the cost of fossil fuels increases because people still need fossil fuel energy to heat their homes, cook, and get to work, not to mention the thousands of products that are made from or made with fossil fuels.

Cap-and-Trade Emissions

Cap-and-trade takes the approach of limiting total greenhouse gas emissions by allowing industries with low emissions to sell their extra allowances to larger emitters. An incentive is provided for industries to reduce their emissions and receive revenue in return for part of their allowances. Overall costs increase, so they will be passed along to consumers in a manner like a carbon tax and cause similar hardships.

Wealth Tax

Taxing the world's wealthiest people could help poorer countries shift to low-carbon economies and recover from climate damage. [219] The logic for this approach appears to embody the suggestion that rich people are responsible for carbon dioxide emissions, so they owe money to poor people as reparations for damage caused. But that logic is flawed, in part because they developed processes and products that directly and indirectly benefited poor people.

As a practical matter, it would seem unfair if a rich software developer whose company had low emissions paid more than a rich chemical, cement, or steel magnate whose company had high emissions. Similarly, it seems unfair that a manufacturer of low-cost cell phones that are accessible to billions of poor people pay reparations. Medicines developed by rich people and sold at significantly lower costs than in rich countries improve the lives of poor people every day. Is it fair for these manufacturers pay reparations?

Further, a wealth tax effectively confiscates money from people who have successfully accumulated wealth and puts it into the hands of people and organizations that have not. This is counterproductive and does not seem like a good idea.

Adaptation

Humans are remarkably adept at adapting to adverse conditions. For example, Eskimos live in extremely cold and remote environments, while Bedouins live in extremely hot and arid locations. Therefore, humans should be able to adapt to most of the effects of climate change.

Floods that result from rising water levels can be mitigated by constructing levees, as was done in New Orleans and The Netherlands, both of which are substantially located below sea level. The effects of forest fires can be mitigated with proper forest maintenance techniques such as removing dead vegetation, maintaining fire breaks, and building structures away from areas subject to fires. The effects of floods near coasts can be mitigated by constructing coastal barriers and not building in areas subject to flooding. The effects of excessive hot and cold temperatures can be mitigated with a heating, ventilation, and air conditioning (HVAC) system.

It is generally easier and more expedient for humans to adapt to nature than to adapt nature to suit humans.

Plant Vegetation

One square kilometer of forest removes between 450 and 4070 tons of carbon dioxide annually for the first 20 years of growth, depending on species. [220] Therefore, planting forests helps reduce the concentration of carbon dioxide in the atmosphere. Conversely, deforestation to install energy sources, such as solar energy farms, negates the benefits of forests.

Planting trees and other vegetation applies universally to all source, electric, and end-use sectors because it can remove carbon dioxide from the atmosphere and hold that carbon until it dies, decays, and releases much of its carbon dioxide back into the atmosphere. This can help reduce the concentration of carbon dioxide in the atmosphere while it is alive but will not return all of the carbon back to the ground from which it came.

161

Planting trees is a worthwhile endeavor. However, Figure 9-9 shows that global carbon dioxide emissions are approximately 37,000,000,000 tons, so its global impact will be small.

Population Reduction

Carbon dioxide emissions would be lower if fewer people inhabited the Earth. War, disease, and famine have caused the global population to decrease for millennia. It is possible but unlikely that catastrophic events will occur that reduce the current global population by any significant amount.

The results of human events, such as the anticipated 600 to 700 million population loss in China by 2100, will offset significant population growth in other parts of the world. Rising standards of living in countries not living at Level 4 will further reduce the global fertility rate and stabilize the global population. Stated differently, prosperity tends to keep the population under control.

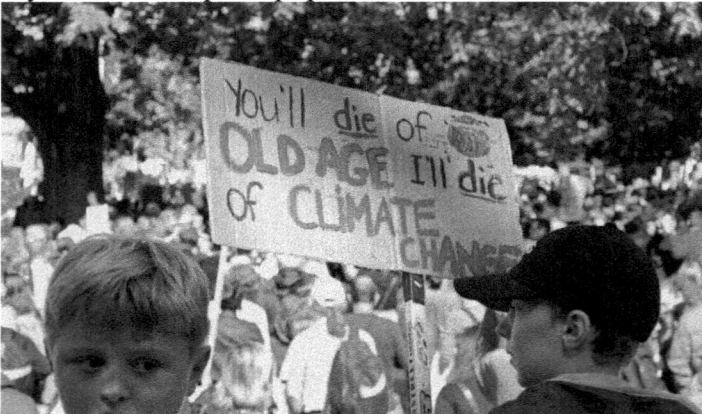

Figure 12-3 Climate Change Protest Sign [221]

More subtle is the indoctrination of children and young adults, some of whom might believe that they will not die from old age but rather from the effects of climate change, as shown in Figure 12-3. This mindset encourages people to have fewer children or no children at all. Although not reflecting reality, these thoughts can disincentivize a healthy lifestyle, reduce ambition, and impart a feeling of hopelessness that can potentially cause individuals to become depressed, drop out of society, or commit suicide. After all, what is the point if humans are on a path to becoming extinct?

Reduce, Reuse, and Recycle

Various governmental agencies have implemented reduce, reuse, and recycle programs to suggest actions that individuals can take to reduce their carbon footprints. These actions can be inconvenient and typically provide relatively small carbon footprint reductions. Nonetheless, many components of these programs do make sense and should be practiced as reasonable.

It would be appropriate to discuss reduce, reuse, and recycle programs as part of the residential and commercial sectors. However, it is discussed here because it applies universally to all source, electric, and end-use sectors.

Reducing means cutting back on the amounts of materials and goods consumed, and the amount of waste generated. Examples include not purchasing more dress shirts if you already have enough, not purchasing a new food processor if the old one still works, not discarding paper towels after light use, and not accelerating excessively when driving. Others include buying food in bulk to reduce packaging waste, turning lights off when not in use, quickly fixing leaking toilets and faucets, using reusable shopping bags, and putting off the purchase of a new vehicle. There are literally hundreds, if not thousands, of actions that can and should be taken to reduce carbon footprints.

Reusing means finding new ways to use things that otherwise would have been thrown out. For example, takeout food containers can be used to store leftovers or to take lunch to work, old clothing can be used as rags, old sheets can be made into handkerchiefs, and items such as computers and clothing can be donated to charitable organizations or community centers.

Recycling involves the collection of materials such as paper, metal, glass, and plastic that are subsequently processed into materials that will later be used to produce goods. Limited recycling programs with collection points and times have been around since at least the 1960s. However, mid-2020s programs typically involve weekly home pickups. Purchasing goods that use recycled materials supports this effort.

The following are some interesting recycling facts:
- Recycling one aluminum can save the equivalent of half a gallon of gasoline.

- The energy saved by recycling one pound of steel can operate a 60-watt light bulb for over a day.
- Approximately 25 million trees would be saved if Americans recycled one-tenth of their newspapers.
- Recycling plastic saves twice as much energy as burning it in an incinerator.
- The energy saved from recycling one glass bottle can run a 100-watt light bulb for four hours, cause 20% less air pollution, and cause 50% less water pollution than when a new bottle is made from raw materials.
- Packaging represents about 65% of household trash.
- Out of every $10 spent buying things, $1 (10%) goes for packaging that is thrown away. [222]

Approximately 75 percent of all aluminum produced in history is still in use, and many countries recycle well over 90 percent of their aluminum cans. [223] Steel is the most recycled material in the world, and 60 percent of it is recycled. [224] Approximately 68 percent of paper was recycled in 2021. [225] Globally, only 9 percent and 19 percent of plastics are recycled or incinerated, respectively. [226] This data suggests that paper and especially plastic recycling should be investigated for improvement, however the wide variety of plastics available provides a significant headwind. [227]

Programs to reduce, reuse, and recycle are beneficial to reduce waste and save energy, and they should be implemented. However, part of their purpose is to reduce carbon dioxide emissions, which is not mentioned. Applying Rosling's *gap instinct* for perspective would reveal that these programs have a small effect on global carbon dioxide emissions, even if they are vigorously implemented.

Reduce Airplane Travel
Limiting the number of trips people can take on airplanes will reduce carbon dioxide emissions. A suggested limit of 4 plane journeys over a lifetime was rather well received in a study, where 41% of French people said that they were in favor. [228] However, France is the 49th largest country in the world, and has good roads and high-speed rail service. However, people living in the largest countries, such as Russia, Canada, China, United States, Brazil, Australia, and India, with long distances over varied terrains would

find 4 trips per lifetime unacceptable, as would people who need to travel for business.

Numerous calls have been made to reduce carbon dioxide emissions by flying on commercial airline flights instead of on private jets that have higher emissions per person per mile.

Near-Zero-Emission Processes

Processes and equipment can be conceptualized as systems that have inputs and outputs, such as a power plant that uses fossil fuel and produces electricity, respectively. In general, conventional processes can produce significant amounts of carbon dioxide. Near-zero-emission processes are designed to operate with close to net zero carbon dioxide emissions. This can be accomplished in several ways, such as consuming carbon dioxide in the process, consuming renewable raw materials, consuming renewable energy, consuming raw materials that have no emissions, removing emissions before discharge, and removing carbon dioxide directly from the atmosphere. Note that many near-zero-emission processes emit carbon dioxide that comes from the atmosphere and not from fossil fuels.

By way of clarification, ancillary processes that have carbon footprints are ignored. For example, ethanol can be made from renewable plants, such as sugar cane and corn, using near-zero-emission processes. Ancillary processes, such as equipment, equipment operation, heating, cooling, transportation, and the like, are ignored because they have relatively small carbon footprints as compared to ethanol production processes that consume fossil fuels.

Some processes can be designed to *consume carbon dioxide in the process*, which can sometimes become carbon negative. For example, the manufacture of fertilizer and carbonated beverages utilizes large amounts of carbon dioxide, as does the oil and gas industry.

Processes that *consume renewable raw materials* have near-zero-emissions with respect to these raw materials. The drive to become more sustainable is pushing industry to increase the use of renewable raw materials. Headwinds to using renewable raw materials include the lack of renewable sources for the complex molecules that are contained in fossil fuels, the reliability of supply, additional handling requirements, and cost.

A process that *consumes renewable energy* has near-zero-emissions with respect to this energy input. However, the process itself can have a significant carbon footprint due to its raw materials and other energy inputs. For example, a manufacturing plant with solar panels that provides near-zero-emissions electricity for the entire plant might consume fossil fuels to produce steam and raw materials derived from fossil fuels, both of which can emit significant amounts of carbon dioxide.

A process that *consumes raw materials that have no emissions* has near-zero-emissions with respect to these raw materials. For example, it was shown in Chapter 6 that burning hydrogen produces water vapor and no carbon dioxide. However, hydrogen production processes directly or indirectly emit carbon dioxide. Ongoing research and development is proceeding to develop viable processes that can be implemented at scale.

A variation on this concept is using raw materials that reduce targeted emissions, such as burning low-sulfur oil in boilers to reduce sulfur emissions into the atmosphere and removing lead from gasoline to eliminate lead emissions.

Another approach is to *remove emissions before discharge* to mitigate emissions. For example, gasoline vehicles use catalytic converters to effectively eliminate emissions of toxic gases and pollutants. Many industrial facilities have air quality control (AQC) and water quality control (WQC) equipment that removes pollutants before discharging air and water, respectively, into the environment. Carbon capture processes can remove carbon dioxide from point sources such as boilers that generate electricity and provide heat.

Direct air capture processes are available that can *remove carbon dioxide directly from the atmosphere* so that it can be permanently stored. These systems have headwinds that include the relatively low efficiency with which carbon dioxide can be purified from ambient air containing only 420 ppm of carbon dioxide. Unlike the previously described processes, direct removal processes can be located virtually anywhere, scaled to remove more carbon dioxide, and operated by independent contractors.

Upgrading or replacing existing processes that consume fossil fuels to enable them to operate as near-zero-emission processes will reduce carbon dioxide emissions. However, there are headwinds because few such processes exist, and many are not feasible.

Research and development efforts continue to pursue making near-zero-emission processes technically and economically viable.

Carbon Capture, Utilization, and Storage (CCUS)

Removing carbon dioxide using direct air capture from the atmosphere and carbon capture from point sources for permanent storage have the potential to reduce the amount of carbon dioxide in the atmosphere and emissions, respectively. The following discussion ignores the carbon footprint of ancillary processes, equipment, and storage.

Direct air capture offers conceptual simplicity and the ability to locate modular processing units virtually anywhere electricity is available. Assuming that approximately 1.2 MWh of electricity is consumed to remove 1 metric ton of carbon dioxide from the atmosphere, [229] generating this amount of electricity from coal and natural gas will emit approximately 1.305 [D] and 0.515 [E] metric tons of carbon dioxide, respectively. Therefore, operating this process when marginal electricity is generated from coal will **increase** emissions by approximately 30 percent, whereas using marginal electricity generated from natural gas would decrease emissions by approximately one-half. However, the benefit of using natural gas will virtually disappear if 2 MWh of electricity is required to perform the same removal, as suggested in Figure 12-4.

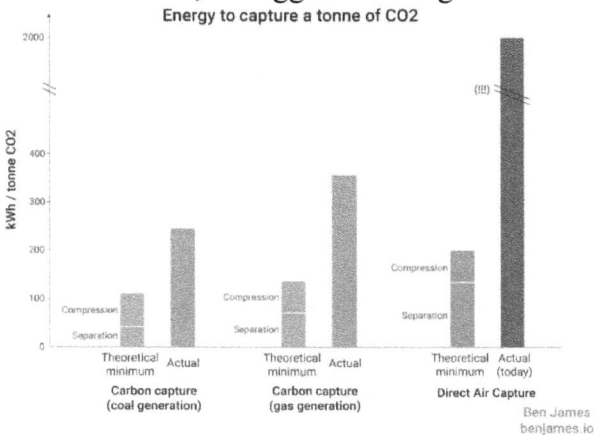

Figure 12-4 – Energy to Capture a Tonne of Carbon Dioxide [230]

[D] 1.2 MWh * 2398 pounds per MWh (from Chapter 7) / 2205 pounds per metric ton

[E] 1.2 MWh * 946 pounds per MWh (from Chapter 7) / 2205 pounds per metric ton

167

Calculations in the previous paragraph suggest that the environmental effects of direct air capture are detrimental, small, or non-existent. Nonetheless, research continues in the quest to improve the efficiency with which carbon dioxide is removed from the atmosphere. It should be noted that direct air capture projects can be economically viable when they are subsidized, or when organizations are willing to pay the price to offset their carbon dioxide emissions.

Operating direct air capture processes using near-zero-emission electricity generation processes can mitigate this issue but begs an investigation into whether more significant emission reductions can be obtained by using it to displace electricity generated from fossil fuels, instead of constructing direct air capture installations.

Carbon capture processes consume less electricity than direct air capture processes, in large part because the streams entering these processes have significantly higher concentrations of carbon dioxide than the atmospheric air that enters direct air capture processes. Per Figure 12-4, carbon capture from point sources is estimated to require approximately 0.25 and 0.35 MWh of electricity to remove 1 metric ton of carbon dioxide using electricity derived from coal and natural gas, respectively. Therefore, approximately 0.27 [F] and 0.15 [G] tons of carbon dioxide would be produced from burning coal and natural gas, respectively, to generate the electricity to remove 1 ton of carbon dioxide. These emissions represent significant improvements when compared to direct air capture processes. Research continues to further reduce electricity and energy requirements.

Utilization of captured carbon dioxide in existing or new processes is inevitable. However, after widespread carbon capture implementation, the amount of carbon dioxide recovered will result in a huge excess of available carbon dioxide. As such, research into *storage* is ongoing to develop processes to permanently store captured carbon dioxide.

Implementation

The way new technologies and practices are implemented and presented to the population is extremely important. It is desirable to

---

[F] 1.305 metric tons * (0.25 MWh / 1.2 MWh)
[G] 0.515 metric tons * (0.35 MWh / 1.2 MWh)

incorporate new technologies seamlessly into daily life without adversely affecting people, and preferably without their knowledge.

For example, the introduction of LED light bulbs addressed carbon emissions, reduced electricity demand, and did not adversely affect people's lives. LED light bulbs look and operate like their incandescent predecessors, use a fraction of the energy, and last much longer. This change was relatively seamless because the new bulbs did not adversely affect people but rather saved them time and money. Such applications of technology are desirable because they significantly reduce energy consumption and emissions without the population.

An example of change that occurred unbeknownst to most people and without adversely affecting the population was the migration to natural gas combined-cycle power plants in the United States. They are much more efficient than conventional power plants, so they have not only kept electricity costs low, but also significantly reduced the amount of carbon dioxide produced per megawatt hour (MWh) produced. Expenditures to build these combined-cycle power plants were justified by profit while also providing a significant environmental benefit. This transition occurred within the electric power sector, so end-users did not even know that it occurred and were not inconvenienced.

Not all actions will be desirable, and some might be inconvenient, such as recycling plastic, metal, and paper.

Actions such as levying exorbitant fossil fuel taxes would be financially challenging for many people, while immediately stopping the burning of all fossil fuels would be intolerable for virtually the entire population. Therefore, changes should be targeted and carefully implemented so that the population can continue to move forward, preferably unimpaired, and unimpeded. If this is not possible, extreme care should be taken to minimize inconvenience and hardship. In either case, efforts should be made to fully understand the positive and negative effects of the changes and mitigate the negative effects.

Not all actions have positive outcomes, even when well-intentioned and seemingly logical. For example, various jurisdictions in the United States banned the use of single-use plastic bags in the early 2020s to reduce waste, reduce the consumption of fossil fuels to make bags, and address environmental concerns. This mandate irritated customers but resulted in a 60% drop in bag sales in New Jersey in 2022. [231] This sounds like a success story.

However, the reusable replacement bags manufactured using little recycled material require 6 times more plastic, so nearly 3 times more plastic was consumed for bags. Greenhouse gas emissions increased by 500 percent, and it was found that a typical grocery retailer location could profit by approximately USD 200,000 per year from alternative bag sales. [232] In other words, aside from irritating customers, this ban dramatically increased plastic production, consumption of fossil fuels, greenhouse gas emissions, and costs to the consumer. **Logical and seemingly well-reasoned actions can have counterproductive results, especially when the population is involved.** That said, it is possible that this increase could be temporary, because customers might become so annoyed with their stockpile of plastic bags that they use them as intended and reduce plastic consumption.

Global Approach

It is pragmatic to understand and accept that it will take many decades to mitigate the changes that have occurred in the Earth's atmosphere since around 1910. However, it is inevitable that future technological innovations will prove beneficial in reducing carbon dioxide in the atmosphere. It is important to understand that technology that becomes viable in the future may not resemble currently available technology.

Discontinuing the burning of fossil fuels, increasing the cost of fossil fuels, implementing cap-and-trade emissions, and imposing a wealth tax are counterproductive and should be discouraged. Adaptation, planting trees, adding vegetation, and reduce, reuse, and recycle programs are important and should be pursued. However, even if everyone does them, their effect is dwarfed by the magnitude of global emissions, so they cannot effectively mitigate the growing concentration of carbon dioxide in the atmosphere. Stated differently, **nothing that is being done in the mid-2020s will seriously mitigate carbon dioxide emissions**, so further implementation of these marginal strategies should be carefully evaluated for their efficacy as compared with developing near-zero-emission processes.

That said, technological innovation should be vigorously pursued to develop efficient, near-zero-emission processes and energy sources. Implementation has the potential to drastically reduce *net carbon dioxide* emissions and **enable the carbon dioxide in the atmosphere to be controlled at a desired concentration to best serve the Earth and its inhabitants**.

The suggested approach to addressing fossil fuel processes is to examine the relationships between fuel sources and end-uses to identify strategies and processes that can significantly reduce carbon dioxide emissions. Identifying these strategies will enable researchers to better focus their efforts. It should be clearly understood that some technologies will not be available yet, while others may not be efficient. Implementation of these technologies should be delayed until they are viable, lest resources be diverted away from more promising projects.

Engineers regularly address big problems that appear to be unsolvable by breaking them down into smaller problems that can be solved. Carbon dioxide emissions are no different. End-users drive energy consumption, so analysis in subsequent chapters focuses on addressing emissions in each sector of the economy. Particular attention is paid to each sector's energy inputs, which are located on the left side of each sector box in Figure 8-1.

Chapter 13: Electricity Generation Sector

If it weren't for electricity, we'd all be watching television by candlelight. --- George Gobel

Snowstorms are common for those living far from the equator, but not much beats the two meters of snow that dropped in areas near Buffalo, New York, in December 2022. One can only imagine being marooned at home by impassible streets and sidewalks. Things can get eerily quiet when there is no movement.

In contrast, steel mills are filled with loud noises made by large equipment. For example, dumping 50 tons of scrap metal into a steel-making furnace is nothing short of loud. Adding another 150 tons of molten steel is a bit quieter, but still loud. The building shakes when the scrap bin and ladle are raised and bumped into the vessel to remove as much material as possible. Even the smaller equipment makes a considerable amount of noise.

Imagine my arriving at work one morning and finding this huge steel mill totally silent. Nothing moved. No sounds. Nothing. Nada. Word had it that there was a problem at the power plant. Electricity was restored a couple of hours later, and the noise started again.

Now imagine what your life would be like if the electricity went out for months. Your world would become quiet, but your primary concern would be survival.

---XXX---

Figure 8-1 shows that the largest energy user in the United States is the electric power sector, which consumes approximately 38 percent (36.7/97.3) of all energy sources combined. A whopping 65 percent of the energy consumed by the electric power sector, or approximately 24 percent (23.8/97.3) of all energy sources, is wasted. The existence of this wasted energy represents a huge opportunity to not only improve the efficiency of the electric power sector but also significantly reduce emissions.

Figure 8-1 also shows that fossil fuels account for approximately 59 percent (1+32+26) of the energy consumed to generate electricity. Therefore, transitioning electric power generation from fossil fuel sources to energy sources and processes that do not emit carbon dioxide would eliminate a significant amount of carbon dioxide emissions. However, there are headwinds to

172

implementation, including the likelihood that the transition would entail replacing over half of the existing electrical generation capacity in the United States.

Electric power plants have carbon footprints associated with ancillary processes that involve equipment, buildings, structures, space heating, air conditioning equipment, lubricating oil, transportation vehicles, and the like. The discussion below regarding carbon dioxide emissions specifically refers to the operation of the power plant processes themselves and not their ancillary processes.

The design life (years), average retirement age between 2000 and 2015 (years), average age (years), generated power (percent annually), and generating capacity in gigawatts (GW) in the United States in the early 2020s are tabulated in Table 13-1, by fuel source, for power plants with at least one megawatt of total electricity generating capacity. The United States generated 4,243,000 GWh of electricity in 2022. [233]

Fuel Source	Design Life	Retirement Age	Average Age	Generation (%)	Capacity (GW)
Coal	40	54	39	26	210
Oil	30-50	38	---	1	36
Natural Gas	30	38	22	32	491
Nuclear	50	35	36	22	95
Hydro	25	70	64	6	103
Geothermal	30	---	---	0.4	2.6
Solar	30	---	---	3.4	61 *
Wind	20	20	---	10	132 *
				Total	1130.6

* Intermittent power generation

Table 13-1 Summary of Power Generation Plants in the United States [234]

Coal

The bottom line in the coal source box in Figure 8-1 indicates that 90 percent of the coal produced is burned to generate electricity. As previously discussed, the carbon footprint of coal-fired power plants is approximately 2.54 times that of a gas-fired power plant for the same amount of electricity. Numerous power plants that burn coal have either been converted to natural gas or replaced with more fuel-efficient combined-cycle power plants that burn natural gas.

The average age and retirement age of coal-fired plants in the United States are 39 and 54 years, respectively. Coal-fired plants are aging, and approximately half of the remaining coal-burning capacity in the United States is planned for retirement prior to around 2040. This will reduce carbon dioxide emissions, assuming that they are replaced with plants that burn other fuels or use technologies that do not emit carbon dioxide.

Given the large carbon footprint associated with electricity generated in coal-fired plants, it would not be preferred to build or operate coal-fired plants. To wit, many coal-fired plants are aging and are being retired or replaced. India has pledged not to build more coal-fired generating plants. [235] China pledged not to back the building of coal-fired plants outside of China. [236] However, the construction of over 100 GW of new coal-fired power plants in China was approved in 2022, which was six times as large as that in the rest of the world combined [237] and approximately 10 percent of China's coal plant capacity. Except for China, these actions illustrate the declining consumption of coal as a fuel source, as is evident in Figure 6-1.

However, research continues to develop technology to convert coal into liquid synfuels such as diesel, gasoline, and jet fuel. However, the emissions associated with these processes will not be known until they are fully developed.

Oil

Figure 8-1 shows that burning oil accounts for only 1 percent of the electricity generated. This is not surprising given that increased production of natural gas as a byproduct of shale oil production has made natural gas less expensive than oil on a heating value basis in North America. By way of operational flexibility, it is relatively straightforward to convert an existing oil-burning power plant to natural gas, even if it was not so designed when constructed.

Natural Gas

Figure 8-1 shows that natural gas accounts for approximately 32 percent of the energy consumed by the electric power sector. Conventional natural gas fired power plants emit less carbon dioxide to generate the same amount of electricity as plants that burn coal or oil.

Combined-cycle natural gas power plants make significantly more efficient use of fuel and emit lower carbon dioxide emissions per unit of electricity generated as compared to traditional natural gas generation plants. This makes combined-cycle natural gas power plants strongly preferred in new and converted fossil fuel power plant projects. Existing combined-cycle installations have been operating for a relatively short period of time and will continue to operate for decades.

Oxyfuel Combustion

Fossil fuels can be combusted using oxygen instead of air to reduce energy consumption because no nitrogen needs to be heated, thereby reducing heat losses out of the stack. However, energy is required to produce oxygen and must be considered. It is noteworthy that the products of combustion are only carbon dioxide and water vapor.

Near-Zero-Emission Fossil Fuel Combustion Processes

Carbon dioxide is produced by the combustion of fossil fuels and renewable fuels derived from plants and animals. Nonetheless, these processes can be designed to be near-zero-emission combustion processes by burning fuel that contains little or no carbon or by removing carbon dioxide from their products of combustion.

For example, hydrogen and ammonia do not contain carbon, so carbon dioxide is not emitted when they are burned or otherwise transformed to produce electricity. Hydrogen and ammonia are not energy sources but rather are produced using electricity that is largely generated by burning fossil fuels. Generating electricity from hydrogen appears attractive. However, its headwinds include its inefficient production, high cost, high storage pressure, flammability, safety concerns, and the difficulty with which it is transported. Nonetheless, utilizing hydrogen or ammonia to generate electricity could become viable if efficient production and electricity generation processes are developed.

Products of combustion typically contain over 10 percent carbon dioxide, so removal is considerably easier than direct removal from the atmosphere, where its concentration is approximately 0.042 percent (420 ppm). Carbon capture, utilization, and storage systems (CCUS) that are being developed can remove most of the carbon dioxide from products of combustion for utilization elsewhere, such as in the production of a useful product, or for reaction and incorporation into a solid that can be permanently stored. This technology has the potential to transform both new and existing fossil fuel combustion processes that emit large amounts of carbon dioxide into near-zero-emission processes. Headwinds include low process efficiency, equipment costs, and operating costs. Importantly, this technology can be **implemented within the electric sector, which will not affect the population** on a day-to-day basis, as does recycling, thermostat adjustments, and the like.

Another approach is to use oxyfuel. By emitting only carbon dioxide and water vapor, the carbon dioxide can be isolated by reducing its temperature to condense the water vapor.

Chapter 12 concluded that **nothing that is being done in the mid-2020s will seriously mitigate carbon dioxide emissions**, so further implementation of marginal strategies should be carefully evaluated for their efficacy as compared with developing near-zero-emission processes. Therefore, in the absence of a widely implemented scalable near-zero-emission energy source that generates reliable electricity without consuming fossil fuels, the **implementation of efficient carbon dioxide removal technologies that transform fossil fuel processes into near-zero-emission processes is fundamental to effectively stabilizing and then lowering the concentration of carbon dioxide in the atmosphere**.

Information presented in Chapter 10 revealed that it will take approximately 251 years for the concentration of carbon dioxide in the atmosphere to reach 1000 ppm at its current rate of increase. Chapter 9 showed that if carbon dioxide emissions were immediately reduced to zero, it would take considerably longer than 52 years to return the atmosphere to a concentration of 300 ppm.

Further, based on work by Derek de Solla Price, Eric Gastfriend estimated that 90 percent of all the scientists that ever lived were still alive in 2015. This growth in science "has taken place almost entirely in the past 200 years - one tenth of one percent of our species' 200,000-year history. Never before have we had so many people whose sole purpose of work is to better understand how the world works. [238] Given the amount of manpower and technology available, it is reasonable to presume that the scientific community will discover ways to efficiently produce carbon-free fuel, develop efficient carbon capture, utilization, and storage systems (CCUS), or develop other processes that can stabilize or reduce the concentration of carbon dioxide in the atmosphere decades before the effects of carbon dioxide in the atmosphere become critical.

Therefore, there is time to wait for the development of efficient carbon capture, utilization, and storage systems (CCUS) or another technology that has not yet been identified. Stated differently, there is no need to act hastily to mitigate emissions. **The real danger to the population is implementing mitigation systems too soon**, which will result in high equipment and installation costs, high operating costs, and wasted fuel, as compared to installing more efficient systems that become available as the technology develops and matures.

Carbon dioxide is produced by the combustion of renewable plants and animal fuels to produce electricity. However, no *net carbon dioxide* is produced because renewable fuels are derived from carbon in the atmosphere. Interestingly, adding carbon capture, utilization, and storage systems (CCUS) to a process that burns renewable fuels effectively transforms it into a carbon-negative process that removes carbon dioxide from the atmosphere.

The following sections discuss electrical power generation processes that do not consume fossil fuels.

Nuclear

Nuclear power plants do not emit carbon dioxide and therefore meet the more important goal of zero carbon dioxide emissions. Per Figure 8-1, nuclear fuel accounts for approximately 22 percent of the energy consumed to generate electricity.

Nuclear accidents such as those at Three Mile Island (1979), Chernobyl (1986), and Fukushima (2011) have dampened attitudes regarding the construction and operation of nuclear power generating stations. To wit, the most recent nuclear plants in the United States began operation in 1996 and 2016.

Small modular nuclear reactors (SMR) are an emerging class of nuclear fission technology that uses reactors that are smaller than conventional nuclear reactors and that can be built in one location, shipped, assembled, commissioned, and operated at another location. They are expected to have shorter approval cycles and be easier to fund than conventional nuclear reactors. Small nuclear reactors are safer than conventional nuclear reactors because they contain less nuclear fuel and have simpler designs in which fewer things can go wrong.

Small modular nuclear reactors have a small footprint and flexible siting requirements, so they can be located as a stand-alone installation or near existing retiring fossil fuel-fired plants, where they can make use of portions of the existing generating equipment and transmission lines. Using small modular nuclear reactors technology to replace retiring fossil fuel-fired power plants meets the important goal of eliminating a large amount of carbon dioxide emissions. Small nuclear reactors are scheduled to be operational in a chemical plant by 2030. [239]

Figure 13-1 Mini Nuclear Power Plant [240]

Mini nuclear power plants that generate less than 20 MW fit and into an ISO container, such as shown in Figure 13-1, are under development. [241]

Addressing safety concerns is an important consideration for nuclear energy reactor design and operation. Small modular nuclear reactors and mini nuclear power plants contain much smaller amounts of radioactive materials as compared to traditional nuclear power plants, which reduces the complexity of safety systems. One manufacturer boasts that its nuclear reactors are transportable over the existing road and rail infrastructures. They use only four operator-controlled variables, which are the control rods, helium circulator, feedwater pump and turbine throttle valve. Opportunities for operator error are drastically reduced, and automated operations allow for fewer personnel on site. Additionally, fuel is designed not to melt and can withstand extreme temperatures. [242]

Some people are against nuclear power because they consider nuclear power plants to be potential nuclear bombs that cannot be fully protected from cyberattacks and believe this risk will worsen as their sizes are reduced and their numbers multiply. [243] This may or may not prove to be the case. The possibility of terrorism is an additional concern that will need to be addressed.

Waste and Biofuels

Waste-to-energy plants burn trash or garbage, also known as municipal solid waste (MSW), to produce steam that is used to generate electricity and reduce the amount of waste that needs to be sent to a landfill. Waste-to-energy plants are generally considered to produce electricity that is not reliable. Biofuels can sometimes be used to supplement fossil fuels to produce electricity.

Burning waste and biofuel emits *net carbon dioxide* into the atmosphere to the extent that they contain materials derived from fossilized carbon, such as coal, oil, and natural gas. Burning waste and biofuel derived from natural sources such as paper, wood, food, and the like, is carbon neutral.

Hydropower

The energy in flowing water has been used to rotate nearby mechanical equipment for hundreds of years. More recently, natural geological features such as waterfalls have enabled flowing water to be diverted to generate electricity. In addition, the water in reservoirs formed by dams has been used to generate electricity. Hydropower energy is desirable in the sense that electrical production has no carbon footprint. In general, generating electricity from hydropower sources is more reliable than other natural power sources, such as wind and solar.

However, many of the existing natural geological features have already been developed, and building dams to flood large tracts of land is difficult to permit. As such, it appears that many of the potential large hydropower installations have already been built. Nonetheless, hydropower energy projects of all sizes should be pursued to provide power that produces no carbon dioxide emissions.

Tidal power technology can be used to convert energy from the movement of water during the periodic rise and fall of the tides into electrical energy while emitting no carbon dioxide. This emerging technology is not widely practiced but has the potential to generate a significant amount of electricity with no emissions.

Wave energy converters can be used to convert the movement of ocean waves into mechanical or electrical energy. Various attempts have been made to apply this technology.

Geothermal

Geothermal power plants typically use energy from a hot underground reservoir to make steam that is used to generate electricity. They are a reliable source of electricity and do not emit carbon dioxide. The amount of geothermal energy is relatively small because it is only available in a limited number of geological locations.

Solar

Solar panels convert energy from the Sun into electrical energy without emitting carbon dioxide. Electricity production is not reliable because little or no electricity is produced when the solar cells are not in direct sunlight, such as at night or on days with clouds, rain, snow, and the like. Energy storage systems used to improve reliability are discussed below. Solar power installations can typically be expanded by adding more panels, within space limitations.

As stated near Table 13-1, the United States produced 4,243,000 GWh of electricity in 2022. There are 8760 hours in a year, so the hourly average is 484 GW (4,243,000/8760). Per Table 13-1, solar energy accounts for 3.4 percent of the 484 GW hourly average, or 16.46 GW. Given that 61 GW of solar capacity was installed, the solar cells produced electricity at the equivalent amount of full capacity for approximately 27 percent of the time (16.46/61). Therefore, actual solar cell electricity production is approximately one-fourth of its nameplate capacity. Stated differently, it takes on average approximately four times as much installed solar cell capacity to produce the same amount of electricity as would be produced by coal, oil, natural gas, nuclear, hydropower, and geothermal energy sources.

Residential and commercial solar installations are typically located on the roofs of buildings. These locations have the advantages of being close to electrical distribution systems and having a ready support structure for the solar panels. Electricity is typically consumed internally and injected into the electrical grid when the Sun is shining, and supplied by the grid when it is not.

Solar farms are solar electric systems that can contain thousands of solar panels. The land they occupy must be cleared and generally cannot be used for other purposes. Clearing the land will liberate carbon dioxide from the vegetation into the atmosphere and drastically alter the local environment. In addition, removing the vegetation will stop the photosynthesis processes that previously consumed carbon dioxide on the site. Locating a solar farm in a desert will generally provide more sunlight and result in less vegetation loss than other locations. However, the shade from the solar panels will affect the local environment.

The median land use for solar cells is approximately 12 square meters per MWh [244] which is approximately 83,300 MWh per square kilometer per year (1000*1000/12). Figure 6-2 shows that natural gas emits 117 pounds of carbon dioxide per MMBTU, so approximately 15,120 tons of carbon dioxide (3.413*83,300*117/2200) would be emitted to produce the same 83,300 MWh per year. One square kilometer of forest absorbs approximately 1000 tons of carbon dioxide per year. [245] Therefore, generating electricity in this manner provides a net reduction of approximately 14,120 tons of carbon dioxide per square kilometer per year (15,120-1000) as compared to burning natural gas to produce the same amount of electricity.

Table 13-1 shows that 59 percent, or approximately 286 GW of the 484 GW average electricity produced in the United States is derived from fossil fuels. Due to the intermittent nature of solar energy, replacing this generating capacity would require at least four times more installed capacity than the reliable capacity it replaces. Installing 1500 GW of new solar capacity would cover approximately 157,750 square kilometers of land (1500*8760/83.3), which is the approximate size of Montana, the fourth-largest state in the United States. However, Montana is somewhat mountainous, so even more land would be necessary. In other words, solar energy installations take up a large amount of land. Increasing electricity demand would require even more land.

In summary, solar panels are suitable for buildings that receive direct sunlight, and in locations where space and environmental issues are not a concern. However, these concerns do suggest that solar energy alone will not be a practical way to generate all the electricity needed. Improvements in solar cell efficiency may partially mitigate some of these headwinds. Energy storage systems to improve reliability are discussed below.

Wind

Wind turbines can be used to convert energy from the wind into electricity without emitting carbon dioxide or generating heat, but they do not generate reliable electricity because they do not operate when the wind is not blowing. Wind electricity is typically injected into the electric grid when the wind is blowing, and relatively small amounts of electricity are purchased from the grid when the wind is not blowing. Energy storage systems such as batteries can be used to provide more continuous reliability. In principle, more wind turbines can be added to expand wind power installations.

As stated near Table 13-1, the United States produced 4,243,000 GWh of electricity in 2022. There are 8760 hours in a year, so the hourly average is 484 GW (4,243,000/8760). Wind energy accounts for approximately 10 percent of the 484 GW, or 48.4 GW. Given that 132 GW of wind capacity was installed, the wind turbines operated at the equivalent of full capacity for approximately 37 percent of the time (48.4/132). Therefore, actual wind electricity production is approximately one-third of its nameplate capacity. Stated differently, it takes on average approximately three times as much wind turbine installed capacity to replace a given installed capacity of reliable electricity, such as coal, oil, natural gas, nuclear, hydropower, and geothermal energy sources.

Wind turbines have been found to pose threats to birds, make noise, and be unsightly. It is estimated that over 500,000 birds are killed annually. [246] In addition, wind turbines generate noise with documented symptoms such as sleep disturbance, headaches, tinnitus, ear pressure, dizziness, vertigo, nausea, visual blurring, irritability, problems with concentration and memory, and panic episodes. [247] Offshore wind turbines may affect whales and other marine life. [248]

Wind turbine foundations generally utilize small amounts of land that are scattered over wide areas, so the expanses of land between the wind turbines can typically retain their existing use. Energy storage systems to improve reliability are discussed below.

Headwind: Raw Minerals and Disposal

The mining and processing of minerals for the manufacture of solar panels and wind turbines occurs in countries and locations that typically do not have, or do not enforce, environmental controls, health and safety regulations, or labor laws. Mining activities are upstream of solar panel and wind turbine manufacture in many locations worldwide, so it is nearly impossible for final solar panel and wind turbine manufacturers to monitor for compliance. In addition, access to raw materials can be limited by geopolitical issues.

In addition, the electricity output of solar panels diminishes over time, so the panels will eventually need to be replaced. Recycling and disposal of early solar panels and wind turbines need to be adequately addressed so they can be recycled instead of sent to landfills. [249]

The mining of raw materials for wind turbines will be more fully discussed in Chapter 14, in conjunction with the larger issue of the mining of raw materials for battery production.

Headwind: Energy Storage

Energy storage systems, such as batteries, pumped storage hydropower (PSH) systems, and hydrogen storage can be used to provide continuous reliability from energy sources that are not inherently reliable, such as solar and wind, by storing energy when it is not needed and generating electricity when it is needed. Other technologies, such as compressed air, compressed carbon dioxide, thermal, and flywheels, might be feasible in some applications.

Pumped storage hydropower (PSH) systems move water from a low elevation to a higher elevation when electricity is available and allow the water to flow back down and produce electricity during times when demand is high. These systems are viable, but their application is dependent on being located near bodies of water at different elevations, so they cannot be used in most locations.

Energy storage systems do not create energy but rather store it for later use and can incur a significant loss of efficiency. For example, the overall efficiency of a hypothetical pumped storage hydropower (PSH) system with a large pump, motor, water turbine, and generator having efficiencies of 90, 95, 90, and 95 percent, respectively, would be approximately 73 percent $(0.90*0.95*0.90*0.95)$ efficient.

Rechargeable batteries that store electrical energy for later use usually require lithium, cobalt, manganese, nickel, and graphite, some of which are in limited supply from a limited number of sources. The efficiency of battery systems is adversely affected by the process of storing, and later generating electricity when needed. Some batteries, such as rechargeable lithium-ion batteries, have been known to pose a fire hazard. [250] In addition, some of the locations where the raw materials are sourced have questionable labor practices and environmental issues.

Hydrogen storage systems store energy by producing hydrogen when an energy source such as solar or wind is available, and using the hydrogen to produce electricity when the source is not available. Additional development is necessary to increase the efficiency and safety with which hydrogen is produced, and later used to generate electricity. [251]

Compressed air and carbon dioxide [252] systems store energy by pushing air into a salt cavern and later releasing the air through a generator to make electricity. Thermal systems can warm molten salt and then use the heat to generate electricity later. Flywheel systems spin a large mass when electricity is available that can spin a generator later, when electricity is needed.

Headwinds for these systems include size, weight, efficiency, the use of lithium batteries, and the low fuel density and safety concerns inherent with hydrogen. Except for pumped storage hydropower (PSH) systems, energy storage systems are not in general practice at the scale necessary to satisfy demand. However, future development of these and other systems may mitigate some of these headwinds.

Chapter 14: Transportation Sector

The reality about transportation is that it's future oriented.
If we're planning for what we have, we're behind the curve. ---
Anthony Foxx

Consider the following excerpts from an article about the then-upcoming strikes related to transportation and travel.

> ...is a hive of strike action right now, with many employees unhappy that sky-high inflation has not been matched by higher wages. Walkouts are planned all over Europe, showing that it always pays to check before you travel.

> From 27 February to 13 April, they have called for a series of 24-hour walkouts every Monday, Tuesday, and Thursday.

> ...will take part in a nationwide eight-hour strike on 14 April, with the walkout scheduled to start at 9am and end at 5pm.

> Unions have also called for an 12th day of protest action across several sectors - including transport - on 13 April. This is the same date that rubbish collectors ... are set to begin a new rolling strike.

> More than 1,000 members of the ... union including those working in passport offices ... are to walk out from 3 April until 5 May. [253]

It is likely that you have been stuck somewhere far from home, and if not, you have almost certainly been stuck in traffic for longer than you desired. The transportation of people, goods, and services is essential to human activity.

---xxx---

Figure 8-1 shows that the transportation sector in the United States consumes approximately 28 percent (26.9/97.3) of all energy sources combined. Approximately 90 percent of the energy consumed by the transportation sector is derived from petroleum.

Transportation fuels have unique requirements in the sense that they must be readily transported and easily replenished. Therefore, transportation fuels typically have a relatively high energy density, and replenishing stations are available wherever people travel. Fuels such as gasoline and, often, diesel meet these requirements. However, there are a limited number of locations that have replenishing stations for other fuels such as natural gas vehicles (NGV), electric vehicles (EV), and vehicles that consume ethanol.

Transportation Fuels

Gasoline, diesel, and natural gas are fossil fuels used for transportation that have thermal footprints and significant carbon footprints.

Fossil fuels can be transformed into synfuels, which are substitutes for fossil fuels. Examples of synfuels include fuel oil, diesel oil, gasoline, and methanol, made from coal and natural gas. Synfuels have a carbon footprint and a thermal footprint due to their fossil fuel origins.

Biofuels are liquid fuels that are substitutes for fossil fuels but are derived from biomass. Examples of biofuels include ethanol made from corn and sugarcane, and biodiesel, made from vegetable oils, grease, used cooking oils, and animal fats. Biofuels are carbon-neutral and heat-neutral because they are derived from plants, animals, and energy from the Sun. Note that some biofuel production processes utilize waste byproducts as raw materials. However, many biofuels compete with food crops, which can cause food prices to increase and make life more difficult for people living at Level 1 and Level 2, especially when biofuel production is subsidized.

Hydrogen produced from renewable energy sources that have no carbon footprint exhibits no *net carbon dioxide* emissions. However, most hydrogen production involves processes and electricity that are dependent on fossil fuels and thus have a carbon footprint.

It should be noted that end-users determine how much they travel and transport, so they ultimately determine the amount, type, and timing of petroleum products provided to the transportation sector.

Vehicle Technology

Internal combustion engines (ICE) are used to propel conventional vehicles, which are commonly fueled by fossil fuels that produce heat and emit carbon dioxide into the atmosphere.

Hybrid electric vehicles (HEV) generate electricity to operate one or more electric motors that propel the vehicle. Plug-in hybrid electric vehicles (PHEV) are hybrid electric vehicles that have a plug-in connection, allowing them to also be charged from an electrical outlet. Hybrid vehicles commonly run on gasoline, which produces heat and emits carbon dioxide into the atmosphere.

Electric vehicles (EV) use electricity from an electrical outlet or charging station to charge an onboard storage device, such as a battery, to operate one or more electric motors that propel the vehicle. Electric vehicles themselves have no carbon footprint per se. However, determining the actual carbon footprint of an electric vehicle involves tracing the vehicle and its energy back to their sources.

Headwind: Driving Range and Charging

An electric vehicle can only store a fixed amount of energy, so there is a maximum distance that it can be driven without recharging its battery. This is an inconvenience when the electric vehicle is driven for long distances because the charging infrastructure is nowhere near as ubiquitous as that used to fill the gasoline tank in a conventional or hybrid vehicle. In addition, there are compatibility issues including the charger type, charging plug, and the charging speeds of the charger and electric vehicle. [254]

Vehicle Mileage Example

By way of example, the combined city/highway mileage per gallon (mpg) for a conventional model, a hybrid model, and an electric model made by the same manufacturer are 35 mpg, 50 mpg, and 119 mpg-equivalent (3.6 miles per kWh). [255] The mpg-equivalent (mpge) for electric vehicles is the number of miles that vehicle would travel using the same amount of energy that is contained in one gallon of gasoline.

If the electric power sector generated and provided only renewable energy, then the electric model mileage would be 119 mpg-equivalent, and there would be no carbon footprint. However, this is not the case because the electricity needed to operate the electric vehicle needs to be generated elsewhere. Nuclear and renewable energy generators typically operate at maximum capacity, so the marginal megawatt will typically be generated by coal, oil, natural gas, or, more likely, a combination of coal and natural gas.

Previous calculations estimated the overall efficiency with which electricity is generated and distributed for coal and natural gas to be 30.6 and 42.2 percent, respectively. The mpg-equivalent for the electric model based on the energy consumption at the energy source is approximately 36 mpg (0.306*119) and 50 mpg (0.422*119) when the marginal megawatt is generated by coal and natural gas, respectively. Calculations to determine the carbon dioxide emissions for each model are presented in this footnote [H] summarized in the next two paragraphs.

This electric model consuming electricity derived from coal emits approximately the same emissions as the conventional gasoline model (215 versus 224) and significantly more than the hybrid model (214 versus 157). If its electricity is derived from natural gas, the electric model emits less than the hybrid model (117 versus 157). However, the marginal megawatt is derived from both coal and natural gas in near equal amounts in the United States in the mid-2020s, so the electric model traveling 406 miles would emit slightly more carbon dioxide than the hybrid model (165 versus 157).

Therefore, driving what appears to be a "clean energy" electric model using electricity derived from coal emits essentially the same

---

[H] Gasoline contains 123,361 BTU per gallon, [H] so burning 8.11 gallons of gasoline containing 1 MMBTU would propel the conventional gasoline and hybrid models 284 and 406 (8.11*35 and 8.11*50) miles, respectively, while emitting 157 pounds of carbon dioxide each, per Figure 6-2. When driving 406 miles, the conventional gasoline model would emit approximately 224 pounds of carbon dioxide (157*406/284). The equivalent of 8.11 gallons of gasoline (1 MMBTU) would propel the above electric model using electricity derived from coal and natural gas approximately 406 miles (8.11*50), while emitting approximately 215 and 117 pounds of carbon dioxide, respectively, per Figure 6-2. Driving these same 406 miles using electricity derived from coal and natural gas emits approximately 4 and 48 percent (215/224 and 117/224) less carbon dioxide than the conventional gasoline model.

amount of carbon dioxide in operation as the conventional gasoline model it replaces and significantly more carbon dioxide than the hybrid model. If the electric model uses electricity derived from natural gas, it emits less carbon dioxide than the conventional gasoline model and the hybrid. However, if the electric model is powered by a marginal megawatt produced by equal amounts of coal and natural gas, its emissions are slightly higher than those from the hybrid gasoline model.

Note that the above emissions are for vehicle operation, and do not include ethanol additives or the significant carbon footprints of manufacturing the vehicles and their batteries.

Headwind: Raw Materials for Batteries and Wind Turbines

The mining and processing of minerals for the manufacture of electric vehicle batteries occurs in countries and locations that typically do not have or do not enforce environmental controls, health and safety regulations, or labor laws. Mining activities are upstream of battery and wind turbine manufacturing in many locations worldwide, so it is nearly impossible for manufacturers to monitor for compliance. The availability of raw materials can also be affected by geopolitical issues. Actions that promote the production of ethical electric vehicle batteries and wind turbines would help strengthen compliance. Recycling and disposal of early solar panels and wind turbines need to be adequately addressed so they can be recycled instead of sent to landfills. [256]

The environmental toll of manufacturing electric vehicle batteries can be significant, as described in the following excerpt. (**Emphasis added**)

> Looking upstream at the ore grades, one can estimate the typical quantity of rock that must be extracted from the earth and processed to yield the pure minerals need to fabricate a single battery:
> - Lithium brines typically contain less than 0.1% lithium, so that entails some 25,000 pounds of brines to get the 25 pounds of pure lithium.
> - Cobalt ore grades average about 0.1%, thus nearly 30,000 pounds of ore to get 30 pounds of cobalt.

- Nickel ore grades average about 1%, thus about 6,000 pounds of ore to get 60 pounds of nickel.
- Graphite ore is typically 10%, thus about 1,000 pounds per battery to get 100 pounds of graphite.
- Copper at about 0.6% in the ore, thus about 25,000 pounds of ore per battery to get 90 pounds of copper.

In total then, acquiring just these five elements to produce the 1,000-pound EV battery requires **mining about 90,000 pounds of ore.** To properly account for all the earth moved though – which is relevant to the overall environmental footprint, and mining machinery energy use – one needs to estimate the overburden, or the materials dug up to get to the ore. Depending on ore type and location, overburden ranges from about 3 to 20 tons of earth removed to access each ton of ore.

This means that accessing the 90,000 pounds of ore requires digging and moving between 200,000 and over 1,500,000 pounds of earth – **a rough average of more than 500,000 pounds per battery.** [257]

Based on this information, Table 14-1 shows the amounts of the various minerals that would be required to produce batteries for the approximately 80 million vehicles produced worldwide in 2021.

Table 14-1 Minerals Required for 80 Million Electric Vehicle Batteries

	kg/Battery	Minerals Required (million MT)	2022 Mineral Production (million MT) [258]
Lithium	11.4	0.91	0.13
Cobalt	13.6	1.09	0.19
Nickel	27.3	2.19	3.33
Graphite	45.5	3.64	1.3
Copper	40.9	3.28	22

By observation, the amounts of nickel and copper required for electric vehicle batteries would put a strain on existing production.

However, the amounts of lithium, cobalt, and graphite would require multiples of the estimated 2021 production.

Another excerpt states:

> Demand has soared in recent years as carmakers move toward electric vehicles, as many countries... announce a phase-out of combustion-engine cars. In fact, five times more lithium than is mined currently is going to be necessary to meet global climate targets by 2050, according to the World Bank.
>
> But there's one big problem. Obtaining lithium by conventional means takes its own environmental toll, or rather three: carbon emissions, water, and land.
>
> Lithium is currently sourced mainly from hard rock mines ... where the mineral is extracted from open pit mines and then roasted using fossil fuels - leaves scars in the landscape, requires a large amount of water and releases 15 tonnes of CO_2 for every tonne of lithium... [259]

Another excerpt adds:

> Exactly how much CO_2 is emitted in the long process of making a battery can vary a lot depending on which materials are used, how they're sourced, and what energy sources are used in manufacturing. The vast majority of lithium-ion batteries—about 77% of the world's supply—are manufactured in China, where coal is the primary energy source. (Coal emits roughly twice the amount of greenhouse gases as natural gas, another fossil fuel that can be used in high-heat manufacturing.)
>
> For example, the <model> holds an 80-kWh lithium-ion battery. CO_2 emissions for manufacturing that battery would range between 3120 kg (about 3 tons) and 15,680 kg (about 16 tons). [260]

Electric vehicle batteries degrade faster over time in warmer climates and when they get more use. Protection against battery defects is typically addressed with an 8-year and 100,000-mile warrantee. Once out of warranty, the risk of having to pay for a battery replacement is assumed by the owner of the vehicle. In 2023, replacing an electric vehicle battery will cost between USD 5,000 and USD 20,000 based on the pack, size, and manufacturer. [261] Therefore, battery failure or declining battery performance could cause a vehicle to be scrapped even though the remainder of the vehicle is in usable condition.

That said, **we do not know what we do not know**, and it is possible that the commercialization of a new technology, such as lithium-ion-phosphate (LFP) electric utility batteries, may surface that reduces or eliminates the need for minerals like the way that motorized vehicles mitigated the Great Horse Manure Crisis of 1894 in Chapter 3.

Headwind: Life Cycle Emissions

Figure 14-1 presents life cycle emission information from a manufacturer that offers similar gasoline, diesel, and electric models.

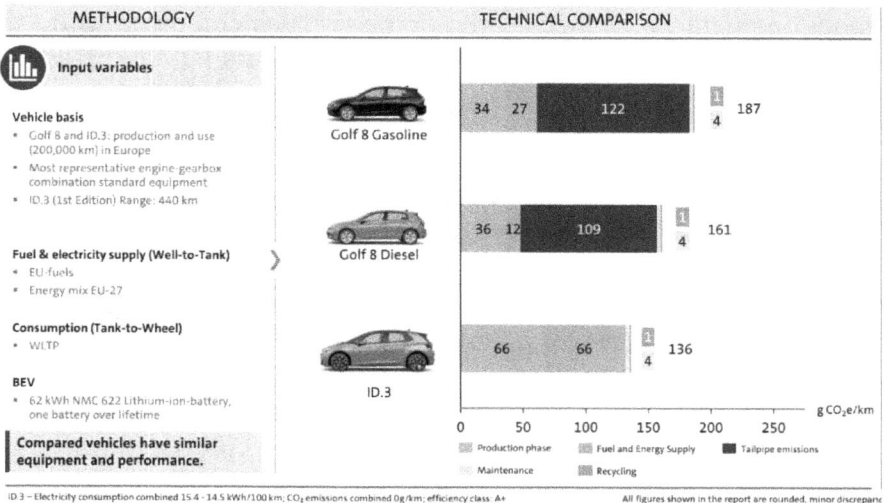

Figure 14-1 Carbon Dioxide Emissions for Gasoline, Diesel, and Electric Models [262]

From observation, the carbon dioxide emitted during the production phase of the electric model is approximately twice that of the gasoline and diesel models. This frontloads the electric model's emissions as compared to the diesel and gasoline models, and notably does not account for the environmental and social effects of mining. A net positive effect on emissions will occur only after driving almost half of the life of the electric model. Electric models retired during the catch-up period would emit more carbon dioxide than diesel or gasoline models.

It should be noted that Figure 14-1 presumes an EU-27 energy mix where approximately one-third of the electricity generated is derived from fossil fuels, as compared to approximately 59 percent in the United States. Therefore, the energy supply emissions for this electric model in the United States would increase by approximately 18, increasing the carbon footprint from 136 to approximately 154.

This electric model driven in the United States over its anticipated life will emit approximately 10 percent {1-(154/161)} and 18 percent {1-(154/187)} less carbon dioxide than diesel and gasoline models, respectively. Electric models that are retired before having driven 200,000 kilometers (124,274 miles) will have a smaller positive or possibly negative effect on emissions, depending on how far they were driven. This analysis shows that this electric model moderately reduces carbon emissions as compared to diesel and gasoline models and that the reduction occurs late in the life of the car.

Hybrid Electric Vehicles (Revisited)

Figure 14-1 presents the total emissions of the specified gasoline, diesel, and electric models. However, it contains a significant error of omission because it does not contain emissions for a hybrid model, even though the manufacturer makes hybrids. Why is that? Might it be that hybrid vehicles and electric vehicles have similar emissions?

In the mileage example above, when the electric vehicle is powered by a marginal megawatt that consists of coal and natural gas, its emissions are slightly higher than those from a hybrid model. The hybrid battery is smaller than the electric vehicle battery and contains less mineral content, so the production phase of the hybrid has a significantly smaller carbon footprint than the electric model. In summary, the hybrid electric model:

- has about the same operating emissions as the electric model.
- has lower emissions in the production phase than the electric model.

- does not frontload load emissions.
- battery has a lower demand for minerals than an electric model, which:
    o is less damaging to the environment.
    o puts less strain on humanitarian abuses.
    o exerts less pricing pressure on limited resources.
    o is less expensive, should replacement become necessary.
- enables continued use of the existing fuel distribution network, mitigating the need to construct a new charging station infrastructure.
- does not have the distance limitations associated with the electric model.

While the hybrid electric model presents numerous advantages, electric models will gradually reduce life cycle emissions as battery technology improves and as electricity generation increasingly shifts to near-zero-emission processes. Therefore, electric models will eventually emit less carbon dioxide than hybrid models.

Chapter 15: Industrial Sector

The Industrial Revolution was another of those extraordinary jumps forward in the story of civilization. --- Stephen Gardiner

The word around the office when I started work at a new company was to stay far away from a certain project that was nearing completion. A plant visit for other meetings revealed that the physical size of the project was much larger than other units in the plant and that a lot of effort was being dedicated to starting it up. But there was history involved and a train wreck to come.

My general understanding was that soon after the project was approved, someone suggested that one of the raw materials could contain oil. Discussions about mitigating this problem resulted in the addition of more equipment. Issues surrounding the additional equipment resulted in the addition of even more equipment. By completion, this multi-million-dollar project had almost tripled in cost.

The good news was that oil was not found in the raw material, and much of the equipment was shut down and not emitting carbon dioxide. The bad news was that the millions of dollars spent to manufacture and install the unnecessary equipment were wasted. The logical thing to do when the question about oil was initially raised would have been to obtain a sample of the raw material and check it for oil. That, apparently, was not done.

---xxx---

This series of events was a blunder, not an error, mistake, or oversight. The industrial sector generally makes few blunders, commits more errors than blunders, makes more mistakes than errors, and commits more oversights than mistakes. Blunders, errors, and mistakes are usually caught and corrected. **The opportunity to reduce emissions often lies in exploiting oversights, especially when their exploitation improves reliability and profitability.**

Exploiting these oversights is typically characterized by customized solutions in many different industries employing many different technologies. Knowing which oversights can be exploited and how to exploit them at a particular industrial site can typically only be done by people working for the company at that site and a few select consultants. This contrasts with the electrical and transportation sectors, which utilize a limited number of processes to make the same product using technologies and plants that are often clones of one another.

Process Improvement

Figure 8-1 shows that the industrial sector in the United States consumes approximately 27 percent (25.9/97.3) of all energy sources combined. Approximately 74 percent (34 + 40) of the energy used by the industrial sector is derived from petroleum and natural gas. Nonetheless, all energy sources are either directly or indirectly important to the industrial sector.

In the electric power sector, the process is remarkably similar in different plants consuming the same fuel. The transportation sector consists of vehicles that are quite similar. The residential and commercial sectors have buildings and houses that are also similar. With some notable exceptions, the industrial sector is comprised of thousands of custom processes that make thousands upon thousands of products.

An incomplete list of industrial sector segments includes agriculture, aircraft, automotive, cement, chemical, energy, food and beverage, glass, hospitals, industrial gases, metallurgy, mining, paper and cellulose, oil and gas, petrochemical, pharmaceutical, ports, shipbuilding, space, steel, water, and wastewater. In the industrial sector, chemical engineering unit operations consist of fluid flow, heat transfer, mass transfer, thermodynamics, and mechanical processes. In addition, most reactions and processes are different, so it is understandable that most chemical plants are custom designed.

The industrial sector is also different in the sense that many petroleum derivatives are not burned but are often used directly in many products such as plastics, fertilizers, medicines, personal care products, fertilizers, and asphalt, most of which would not exist without using petroleum as a raw material. These products have a carbon footprint during extraction, transportation, and processing. However, because they are not burned, they do not emit carbon dioxide, but the eventual disposal of some of these products can produce emissions.

The electrical power sector is regulated so that consumers are protected, and the utilities make a reasonable profit. Transportation sector users purchase fuel when needed at the market price. Residential and commercial users purchase what they need to have heat, electricity, and the like. The industrial sector is somewhat different than the other sectors because it needs to be profitable to survive.

Previous investments in industrial processes and plants run into the trillions of dollars, so making constant changes is not practical. However, resources are applied quickly when an opportunity for improvement is identified because industrial companies constantly strive to increase profitability. In summary, industrial plants tend to have a lot of inertia but can often be nimble at the same time.

Industry has burned and incinerated plant waste for decades to reduce the amount of waste, lower disposal expenses, and utilize its energy content. For example, paper mills, chemical plants, and other operations burn their waste to make steam for use in their processes and sometimes to generate electricity.

Industrial plants are typically comprised of process units that produce intermediate products, final products, and utilities. For example, the reactor unit might convert three chemicals into an intermediate stream containing two intermediate products and waste. An extraction unit might remove the waste so it can be destroyed in an incinerator unit that generates steam for use in the plant. A distillation unit might separate and purify the two intermediate products so they can be made into more products in another reactor unit. Excess steam might be sold to a nearby company.

The overwhelming majority of processes within an industrial plant are complicated. Adding the interconnections between processes and their utilities, such as steam, condensate return, water, cooling water, industrial gases, and electricity, increases the complication. Given the one-of-a-kind nature of most processes, it would not be surprising that only the people working on the process and their consultants would know how to improve or otherwise change and improve the process.

Point Sources

Fortunately, addressing carbon dioxide emissions from complicated industrial plants does not require intimate knowledge of the processes because emissions typically exit the plant at a few fixed locations. These locations are point sources, such as chimneys or stacks, which are often already regulated by a local, state, or national authority.

An industrial plant may have many point sources. However, there may be only a handful that emit carbon dioxide in any appreciable amount, such as from heaters or boilers that use fossil fuel energy sources or in processes that contain or produce carbon dioxide. These point sources, if any, should be identified and their emissions quantified.

Many point sources that emit significant quantities of carbon dioxide in industrial plants are located in the utility plant, where boilers and turbines burn fossil fuels to generate steam and electricity, respectively. However, point sources can be located throughout the plant when warranted. These point sources should likewise be identified, and their emissions quantified.

Near-Zero-Emission Processes

Most point sources contain products of combustion, so transforming these processes into near-zero-emission sources is the similar to approaches suggested in the electric sector, as described in Chapter 13 and not repeated here. **A real danger to the population is implementing mitigation systems too quickly**, which will result in high installation costs, high operating costs, and wasted fuel as compared to installing more efficient systems that will soon be available as technology develops.

Point source emissions that are not the result of combustion processes require special consideration. Depending on location, some processes may lend themselves to using low carbon or renewable fuel sources.

Regardless of how carbon dioxide emissions are mitigated, cogeneration, as described in Chapter 7, should be seriously considered to reduce fuel consumption, save money, and reduce emissions.

Summary

The industrial sector is a collection of disparate processes that need to be operated profitably to survive. However, many of these processes have the common characteristic of containing a limited number of identifiable point sources from which carbon dioxide is emitted. This is similar to the electric sector, but in stark contrast to

199

the transportation sector which has point sources on millions of vehicles.

Chapter 16: Residential and Commercial Sectors

Global warming is something that happens to all of us, all at once. -
-- Larry Brilliant

It was an exciting day when a local heating, ventilating, and air conditioning (HVAC) company came to replace my furnace and air conditioner. It was so exciting that even paying for the furnace could not dampen my spirits because I understood the value that the furnace represented.

The existing conventional natural gas furnaces for the downstairs and upstairs were approximately 45 years old and rated for 80,000 and 105,000 BTU/hour, respectively. I recall that their nameplate efficiency was 80 percent while running in a steady state.

However, these furnaces were designed to turn on when the thermostat signals for more heat and turn off when the house reaches the desired temperature. The house would then cool, and the on-off cycle would repeat. Therefore, the furnace would have to warm up before the fan turned on to blow warm air into the house and cool down afterwards. The warmup and cooldown processes waste energy, so the actual efficiency was probably closer to 65 percent.

The new 90,000 BTU/hour condensing furnace had a 96 percent efficiency rating and was being installed with dampers to direct the warm air flow upstairs, downstairs, or both upstairs and downstairs, depending on the need for heat. Condensing furnaces are designed to utilize not only the heat in the hot products of combustion but also recover heat from the water vapor in the products of combustion when it condenses. You may recall that steam power plants have low overall efficiency because they waste thermal energy when steam is condensed. Condensing furnaces are designed to utilize this energy instead of wasting it.

Producing 90,000 BTU/hour of warm air in the new furnace will consume 93,750 BTU/hour (90,000/0.96) of natural gas, whereas the existing furnace would consume approximately 138,450 BTU/hour (90,000/0.65) of natural gas to produce the same amount of warm air to heat the house. Therefore, natural gas consumption will be reduced by approximately 32 percent {1-(93,750/138,450)}.

Not only that, but the existing furnace pulled its combustion air from the garage, which meant that cold outside air was drawn into the garage. The new furnace pulled in the outside air directly, and the garage became noticeably warmer. Not only that, but the new furnace was also designed to be powered from a plug instead of being hard-wired, which allows connection to a special 12 VDC to 120 VAC inverter connected to a car battery, enabling furnace operation for limited periods of time during power outages without having to purchase a generator. Not only that, but the new air conditioner's motor was also smaller, so it would consume less electricity. Not only that, but the new furnace fan motor was also smaller, so less electricity is needed to circulate air. Not only that, but the fan motor operates at low speed until it needs to go faster, which also saves electricity. Not only that, but the new furnace, which burned less natural gas, emitted less carbon dioxide than the two furnaces it replaced.

With all the positives, why did it take so long for me to replace my furnace? My initial investigations over the years uncovered a limited number of condensing furnaces from little-known manufacturers that appeared to have maintenance and repair issues. What prompted my purchase was a condensing furnace offered by a major manufacturer, and comments attesting to its reliability. For the record, the only problem with my now 8-year-old condensing furnace is that a bug got stuck in the burner after about two months of operation. The combustion chamber has been checked annually since then and has always been reported to be clean.

Condensing furnaces are approximately 10 to 15 percent more expensive than conventional furnaces, depending on size. In 2023, a local installer estimated that 50 percent of homes would install condensing furnaces, while the remainder would install conventional furnaces. Why is this not close to 100% condensing furnaces? After all, reducing gas consumption by approximately 30 percent should sell a condensing furnace on its merits, which include lower carbon emissions.

---xxx---

The residential and commercial sectors are discussed together because the equipment that consumes energy in both segments is similar, albeit with different sizes and operating for different amounts of time.

Figure 8-1 shows that the residential and commercial sectors in the United States consume approximately 21 percent {(11.6+9.1)/97.3} of all energy sources combined. The energy used by the residential and commercial sectors is primarily electricity and natural gas, with some oil consumption.

The residential and commercial sectors are perhaps the most difficult sectors to change, because not only will the population be affected, but help from the population is almost always needed for implementation. Significant help will generally not be forthcoming, even under the best of circumstances. However, proposing changes that adversely affect standards of living almost guarantees no help and, worse yet, vehement opposition. For example, California's mandate to require automakers to provide electric cars was repealed in 1996 after staunch opposition surfaced. [263]

This does not mean that all is lost, but rather that changes should be designed and implemented in a manner that either does not involve the population, or that market forces will promote. For example, a new condensing furnace should sell itself by word-of-mouth education when new owners see dramatic reductions in their gas and electric bills. The reduction in carbon dioxide emissions that occurs due to furnace replacement is transparent to consumers, most of whom will not recognize that it is occurring. Importantly, there should be little or no opposition to installing a more efficient furnace.

Training

The population should have cursory knowledge of the science surrounding global warming, greenhouse gas emissions, and the long-term threat carbon dioxide represents, but this is generally not the case. However, a limited number of people who are close to the intersection between the buyer and seller of equipment that consumes energy should be educated on the operational, economic, and environmental details associated with the equipment being sold.

For appliances that consume a large amount of energy, such as furnaces, manufacturers should train their wholesalers, local salespeople, and installers not only about how to apply and install the equipment but also about the economic benefits of one technology over the other. The latter point should be emphasized: have the wholesalers, local salespeople, and installers sell the savings to the customer as if the furnace is going to be installed in their own house instead of the customer's house. Lower emissions will come along for the ride, unbeknownst to the customer.

In other words, wholesalers, local salespeople, and installers should not focus on making a quick sale where the customer purchases based upon incomplete information, but rather the seller should take the time to show the customer how a particular furnace technology is significantly more economical and, as a side note, has lower emissions. This is idealistic, easier said than done, and against human nature where salespeople make more money by spending less time with individual clients.

To wit, my local contractor has been providing quality furnace installations and repairs for decades. Yet it was obvious to me that there was a good chance that a conventional furnace would have been installed in my house had my request for a condensing furnace not been made because the salesperson's compensation is largely based on the number of furnaces sold, which provides an incentive to spend less time educating me about condensing furnaces and more time pursuing the next sale.

Furnaces

Discussion above makes a strong case for installing a condensing furnace instead of a conventional furnace. Importantly, this discussion shows that the state of technological development is similar to that of LED light bulbs prior to the development of regulations that effectively eliminated the sale of incandescent light bulbs, where (repeating from Chapter 2) legislation or pseudo-legislation transparent to the population, involves a small sector of industry, is feasible, and allows reasonable time for implementation would tend to have better chances for success. Not surprisingly, the United States Department of Energy is finalizing efficiency standards that will effectively prohibit the sale of older furnace designs that are less efficient. [264]

However, it would be also prudent to compare a condensing furnace with an air or heat pump system that might be even more efficient and emit less carbon dioxide, depending on location. Heat pumps operate like an air conditioner by transferring heat from a warm space inside to the environment outside and, in reverse, by heating an inside space when it is cold outside. They operate using electricity and would have a near-zero carbon footprint if the electricity was generated by a near-zero-emission process.

Condensing furnaces utilize almost 100% of the energy in their fuel, whereas heat pumps are often advertised as being hundreds of percent efficient. These claims may be valid for the heat pump itself under certain operating conditions. However, the electricity that operates heat pumps is typically generated with an efficiency of between 30 and 45 percent, so additional calculations are necessary to determine the overall efficiency of the heat pump system as related to its source energy.

Despite my excitement about condensing furnaces, furnaces that operate using only electricity, such as heat pumps, will replace fossil fuel furnaces in many locations due to their high efficiency, good controllability, and lack of need for a natural gas connection.

While installing a heat pump should be considered when purchasing a new or replacement HVAC system, beware of claims that may be true, but do not make a valid comparison. For example, a well-known manufacturer's website stated that "under ideal conditions, a heat pump can transfer 300 percent more energy than it consumes, compared to a high-efficiency gas furnace's 95 percent rating." The statement may be technically correct, albeit noting that the stipulation "under ideal conditions" is poorly defined. However, investigation will show that the 300 percent heat pump efficiency is based on its electrical energy input and not the energy source input. If the electricity delivered to the residence is approximately 30 percent efficient relative to the energy source, the efficiency of the heat pump relative to the energy source would drop to approximately 90 percent (0.3*300), which is similar to the efficiency of the high-efficiency natural gas furnace. This is but one example of the Rosling *gap instinct* where it is implied that one group is different than another when, upon closer inspection, there may be little difference between the groups.

Technical advancements in heat pump technology and application might make the installation of heat pumps ubiquitous. If this occurs, it will become prohibitively expensive to install natural gas distribution systems for residential and commercial buildings that use heat pumps because the remaining natural gas devices consume low quantities of natural gas, so there will be a natural transition to consume only electricity. Prohibitions on the purchase and installation of fossil fuel furnaces will hasten this process, but acting too quickly can significantly increase the cost of heating.

Solar Panels

Solar panels can be installed on the roof or grounds of residential and commercial structures to generate renewable electricity.

Stand-alone solar electricity sources are renewable because their consumption does not affect the power distribution grid. However, most solar power sources are connected to the grid, so consuming electricity reduces the amount of electricity that flows to the grid and increases fossil fuel power generation elsewhere by a similar amount.

Appliances

Gas stoves and water heaters will eventually be replaced by electric devices in structures where furnaces operate using electricity because it will become expensive to install infrastructure to provide small amounts of natural gas for cooking and heating water. In other words, gas stoves and water heaters will be phased out naturally as furnaces are transitioned to operate using electricity.

Air conditioners can consume a large percentage of energy and should be selected based on their energy efficiency and anticipated use. For example, it would be prudent to purchase the most efficient air conditioner for a house in southern Florida that will operate all day and night during the summer months. On the other hand, purchasing that same air conditioner for a house in northern Maine, where it might operate for 8 hours a year, does not seem reasonable because it will save only a small amount of energy during the few hours during which it operates.

Other appliances, such as refrigerators, stoves, washing machines, dishwashers, clothes dryers, light bulbs, radios, televisions, and computers, also consume energy. Selecting efficient appliances and turning them off when not in use will reduce energy consumption and lower carbon dioxide emissions.

Manufacturers design their products to operate in the most efficient manner possible to save energy for the consumer and reduce carbon dioxide emissions. Various testing and labeling standards exist that provide standardized energy consumption information for customers to evaluate when making purchases.

That said, the United States is proposing to use efficiency regulations to impose an electrification agenda [265] to speed up the process. The United States Department of Energy has unveiled new standards for a wide variety of appliances, including gas stoves, clothes washers, refrigerators, air conditioners, and dishwashers. Electrification might make sense for some installations. However, in many installations, the result can be higher equipment, retrofit, and installation costs that are not offset by reduced operating costs, aside from not appreciably reducing emissions. [266] Stated differently, one-size-fits-all solutions may not make sense in many installations.

Construction

Standards often dictate that buildings, windows, and doors should be well insulated. Owners should maintain this insulation and add to it where practical. In general, consideration should be given to designing energy-saving devices into the original design, when they can be incorporated most cost-effectively.

Food

Figure 16-1 shows that diets with meat generally have the highest carbon footprints. Vegan and cultured meat products are coming to market that have lower carbon footprints.

Foodprints by Diet Type: t CO_2e/person

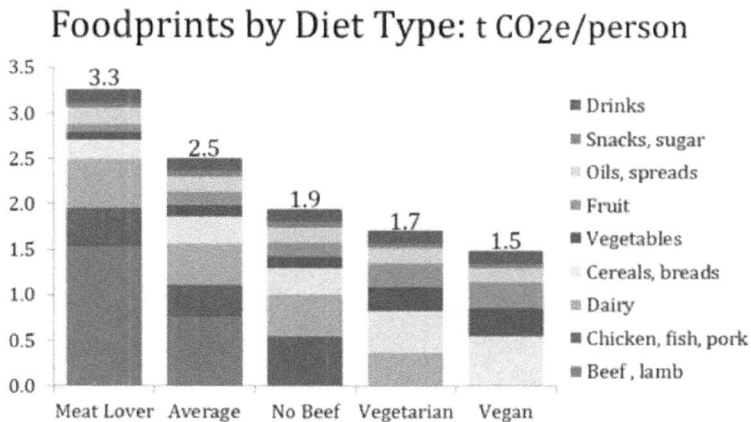

Note: All estimates based on average food production emissions for the US. Footprints include emissions from supply chain losses, consumer waste and consumption.. Each of the four example diets is based on 2,600 kcal of food consumed per day, which in the US equates to around 3,900 kcal of supplied food.

Sources: ERS/USDA, various LCA and EIO-LCA data   Shrink That Footprint

Figure 16-1 Footprints by Diet Type [267]

In addition, organic products will typically have a smaller carbon footprint because fertilizers and other farm products are not derived from petroleum. Shopping locally, purchasing locally grown food, growing your own food, and eating at home reduce transportation footprints.

Relocation

A study in Switzerland determined that the rural carbon footprint is about 20% larger than the urban one, and that wealthier people tend to have larger carbon footprints, regardless of their habitat. [268] Public transportation has a lower carbon footprint than private vehicles. Therefore, the carbon footprint of humans can be reduced by living in an apartment in a city, forfeiting the convenience of having a private car, and getting poorer, virtually all of which will generate opposition from the population.

Another option is to relocate to a country with low emissions and live like the general population to lower carbon emissions. However, this will typically involve learning a new language and living at a lower level, which will likewise not be appealing to many people.

Chapter 17: Emission Reduction Strategy

All the proof of a pudding is in the eating. --- William Camden

Motors are typically designed to perform their designated functions at a single speed. However, sometimes the motor speed can be varied to perform better and more efficiently, such as by using a variable-speed drive. One day, a colleague in the field of flow measurement was specifically interested in the origin of my interest in variable-speed drives and what they had in common with flow measurement.

By way of background, a variable-speed drive varies the speed of a motor that operates a piece of equipment in a process. For example, the motor in a vehicle is a variable-speed drive, where the accelerator pedal is depressed to increase the speed of the vehicle and released to slow it. In industrial plant applications, a variable-speed drive system typically consists of a drive, motor, pump or compressor, and the process within which it operates. Selecting these components requires electrical, electromechanical, mechanical and hydraulics, and process expertise, respectively. Instrumentation and control expertise are required to ensure that the components are properly integrated to function seamlessly in the process. These areas of expertise can be used to calculate the energy savings associated with entire variable-speed drive systems and to justify the purchase and installation of the variable-speed drives.

My immediate answers to my colleague's inquiry were that combining seemingly unrelated technologies interests me and that hydraulics is the common thread between flowmeters and variable-speed drives, respectively. Further, it is of paramount importance to making the process energy-efficient and easy to operate at the same time.

---xxx---

Developing a global emission reduction strategy is not a simple task. The first order of business is to define the problem at hand, which is the increasing carbon dioxide in the atmosphere, and not global warming which, based on what appears to be credible data and analysis, might not even exist.

This chapter utilizes information in prior chapters to develop approaches that can be used to mitigate the accumulation of carbon dioxide in the atmosphere, the efficacy of which is quantified in the next chapter.

Spoiler Alert: Some potential actions may be unpleasant and appear counterintuitive.

The increasing concentration of carbon dioxide in the atmosphere is a global issue that would have already been resolved if there were simple solutions that appealed to everyone. Such is not the case, but the good news is that a path does exist to address the issue. Another piece of good news is that the path involves direct changes to the electric and industrial sectors that can be implemented without directly involving the population. The bad news is that the changes cost money and require technology that is still in development. On a positive note, if appropriately implemented, these changes will indirectly affect the population in only minor ways over time.

Figure 8-1 illustrates some of the many interrelationships between the various sectors, behind which there may be many incongruent occurrences. For example, an industrial plant might generate its own electricity and pay the electric sector to transport its excess electricity over the power grid to another industrial plant, thereby competing with the power sector. In the residential sector, electricity and natural gas providers have programs to help educate consumers on how to reduce their electric and natural gas consumption, so the better they educate consumers, the less energy they sell.

The carbon dioxide concentration in the atmosphere took over two centuries to reach its current concentration. Returning to pre-1800 concentrations will not occur quickly. Nonetheless, there are various paths to effectively eliminating nearly all the carbon dioxide produced by fossil fuels in a targeted, prescriptive manner that does not affect the population.

The activities suggested below will almost certainly be modified as existing technology develops and new technologies emerge. Nonetheless, they present a roadmap forward to address carbon dioxide emissions while reasonably protecting investments in existing installations.

Electricity Generation

The long-term goal is to generate electricity from fuels using near-zero-emission combustion processes. Energy sources that inherently meet this criterion are hydropower, geothermal, tidal, wave, solar, wind, and nuclear. The capacity and location of conventional hydropower and geothermal are limited. Tidal and wave technologies are not fully developed. Solar and wind are not reliable. **Nuclear is the only energy source currently available that can generate zero-emission electricity at the scale necessary to satisfy demand.**

Short-term activities to reduce carbon dioxide emissions can be implemented relatively quickly using existing technology.

- Prioritize the operation of existing power generation equipment to utilize energy sources with the lowest carbon footprints, such as upgrading, expanding, and building hydropower and geothermal plants.
- Until near-zero-emission fossil fuel electricity plants are fully developed:
    - install nuclear or combined-cycle natural gas generation for reliable electricity.
    - install solar or wind generation capacity for unreliable electricity based on their merits **without subsidies**.

Medium-term activities to reduce carbon dioxide emissions generally take longer to implement and utilize technology that is not yet fully developed.

- Legislate and enforce environmental regulations that limit carbon dioxide emissions from fossil fuel power plants to effectively transform them into near-zero-emission processes and allow new and existing power plants to operate using the fuel of their choice, albeit subject to emission limits.
- Develop and install small modular nuclear reactors to generate reliable, zero-emission electricity.

Long-term activities to reduce carbon dioxide emissions take much longer to implement and often utilize technology that may be theoretical or in the research phase.

- Develop nuclear fusion reactors that will be safe, reliable, and have zero emissions.

- Develop other, yet undefined, technologies to generate electricity.

Importantly, these activities can be accomplished within the electric sector without involving the population or relying on their compliance. Have no illusions about the time frame involved, because developing technology, passing legislation, and implementing regulations will take decades.

Point Source Regulation

Huge amounts of investment were made to develop products and build industrial plants to manufacture them. Similarly, large investments were made to install electricity generating plants and commercial buildings.

Effective treatment at point sources will, by its nature, reduce emissions. Further, it will also ensure that the benefiting parties pay for the benefits they receive. This is similar to a carbon tax, where the treatment cost is similarly paid by those who benefit, and not by the population. Understandably, treatment costs will increase product prices and adversely affect economic activity.

Addressing emissions from the point sources in need of remediation can involve large amounts of capital for little benefit or may be difficult to implement due to a lack of space in which to install additional equipment. For example, regulations almost forced an industrial plant to build a multimillion-dollar pollution abatement system to remove a small amount of metal that sloughed off its stainless-steel pipes.

Emissions that require remediation should be regulated for the plant in total, not individual point sources. For example, assume that an electricity plant has four identical point sources and has plans to add one more. If the local authority mandates remediation of 20 percent of its emissions by a certain date, it may be better to incorporate that reduction into remediating the entire new point source instead of partially retrofitting all five of the point sources to accomplish the same reduction. Further, regulations should be based on *total* carbon dioxide emissions per unit of electricity production on an annual basis.

Stated differently, this approach suggests prescriptive solutions that drive the desired result, so the best possible decision about how to achieve that result can be made. In addition, the regulations should apply to new plants, be phased in over time in existing plants, and not be rigorously applied to plants nearing retirement.

International Regulation

Regulating carbon dioxide emissions is a global issue that no one country can effectively address. For example, the country with the most emissions in 2021 was China, at approximately 30.9 percent of global carbon dioxide emissions. [269] Even if China completely stopped emitting, approximately 69.1 percent of global emissions would remain. The next four largest emitters produced 13.5, 7.3, 4.7, and 2.9 percent, with the remaining countries producing less than 2 percent each. Therefore, effective solutions should be embraced globally.

The efficacy of environmental regulations and their level of enforcement vary with location and produce varying results. Imposing strict environmental regulations and enforcement on a mutually agreeable timetable in countries living at Level 4 would seem reasonable. However, other countries will find implementation at the same pace difficult.

Despite being unlikely, it would seem reasonable for the other countries to partially implement regulations, not based on negotiations but rather on a global sliding scale (not groups) predicated on an accepted metric, such as its per capita gross domestic product (corrected for inflation), that is formally published by an agreed-upon organization, such as the International Monetary Fund, World Bank, or United Nations.

Transportation

The overall goal is to transition the transportation sector towards zero-emission processes and fuels, such as hydrogen, renewable fuels such as biofuels, and energy from near-zero-emission power sources.

Being environmentally pragmatic, hybrid electric vehicles that have a smaller carbon footprint than gasoline and electric vehicles that utilize the existing refueling infrastructure are currently preferred but will eventually be phased out naturally in favor of electric vehicles. This will occur when technology advances to the point where electric vehicle emissions have a smaller carbon footprint than hybrids, battery technology is more fully developed, widespread recharging infrastructure is in place, and more reliable near-zero-emission electricity generation capacity is in place.

Large amounts of new electric generating capacity will be needed to accommodate electric vehicles. The average number of miles driven annually per vehicle in the United States is approximately 14,500. [270] The electric model described above would require approximately 4028 kWh (14,500/3.6) of electricity annually. There were approximately 282 million registered motor vehicles in the United States in 2021. [271] If all these were electric vehicles, the transportation sector would consume an additional 1.14 million GW (4028*282,000,000/1,000,000) per year from the electric power sector. For this to occur, the United States would have to generate approximately twice as much electricity as it currently generates. A portion of this electricity could be generated using existing capacity during off-peak hours, creating a need for large-scale energy storage.

Therefore, hybrid electric vehicles present a short- and medium-term opportunity to reduce transportation fuel consumption and moderately reduce carbon dioxide emissions until sufficient reliable near-zero-emission electricity capacity is installed.

Salespeople and potential vehicle purchasers should be educated about both the economic and environmental impacts of the vehicles they purchase so that they sell themselves on their merits, where no mandates should be imposed, and no subsidies should be offered, which will enable technology to be adopted naturally and not create distortions in the market.

An incomplete list of short-term strategies that should be continued and emphasized includes programs to improve public transportation, incentivize reduced vehicle use, perform bio-fuel research, improve mileage standards, research better battery technology, research solar-powered vehicles, and the like.

Industrial

Overall, change will come slowly in the industrial sector. Complete elimination of fossil fuel input might be possible in a relatively small number of processes. However, a realistic goal is to reduce carbon dioxide emissions from industrial processes somewhat by making incremental process changes. Legislation should regulate industrial point sources in a manner similar to those in the electrical sector.

Industrial processes are not changed often, but they can change quickly under the right circumstances. If possible, industrial sector end-users should consider reducing the carbon footprint of processes during the design phase because they can sometimes be designed to accommodate materials from non-fossil fuel sources.

Due to the diverse nature of industrial processes, it is difficult to make suggestions as to how to significantly reduce process waste related to carbon dioxide emissions. For example, power plants produce high-pressure steam, generate electricity by passing it through a steam turbine, and reject waste heat to the environment when the steam condenses. Per Figure 8-1, over half of the energy input to power plants is wasted. Many process plants require a significant amount of steam, which is similar to the type of steam that would be wasted in a power plant. Process plants could be located near power plants to utilize this steam, but this is usually not feasible. However, another alternative that should be seriously investigated is for industrial plants to cogenerate electricity and steam to make almost complete use of the fuel that is burned.

Per Figure 8-1, the industrial sector consumes approximately 9 percent of the coal produced. Much of that coal is consumed to make steel. Electric arc furnaces that do not use coke have been around for decades, but technology is evolving to develop processes that use other fuels, such as hydrogen, to make steel without fossil fuels.

Controlling the speed of a piece of equipment, such as a pump, to match its load can often be accomplished by using a variable-speed drive to reduce the speed of its motor. Operating the motor at a slower speed will decrease electricity consumption, thereby saving energy and reducing the emission of carbon dioxide into the atmosphere. [272]

Biogas produced on farms and other facilities handling agricultural waste can be used to produce heat and generate electricity. In addition, harvesting biogas instead of letting it disperse into the atmosphere reduces methane, carbon dioxide, and nitrous oxide emissions. On a side note, food additives are being developed to reduce methane emissions from belching bovines.

With few exceptions, the custom nature of most industrial processes virtually forces most sites to develop their own customized strategy to reduce emissions. While there may be common paths to eventually generate electricity and provide transportation fuels almost entirely from near-zero-emission energy sources, the industrial sector is virtually forced to address each process individually. Except for point sources of carbon dioxide emission, governmental regulations are of limited benefit in the quest to reduce carbon dioxide emissions.

Another approach to reducing carbon dioxide emissions is to develop economically viable processes that utilize carbon dioxide as a raw material, which will enable captured carbon dioxide to be consumed instead of emitted.

Trade magazines regularly publish process and carbon capture developments that may (or may not) become viable.

- Process to transform carbon dioxide into propane [273]
- Catalyst research focused on addressing plastic waste [274]
- An electrolyzer system that makes propane from carbon dioxide [275]
- Construction of a demonstration plant to convert carbon dioxide to fuels [276]
- Construction of a carbon capture pilot plant [277]
- Construction of a chemical looping combustion demonstration unit [278]
- Electrolysis process that uses seawater to capture and store carbon dioxide [279]
- Using waste plastic to make graphene and hydrogen [280]
- Methane pyrolysis process to produce hydrogen [281]

Residential and Commercial

Most of the energy consumed by the residential and commercial sectors is in the form of electricity and natural gas.

When the electric sector implements the suggestions above, it will largely generate near-zero-emission electricity. Therefore, the residential and commercial sectors will effectively have no carbon footprint from the electricity they consume. With near-zero-emission electricity, heat pumps will replace natural gas furnaces, so the evolution to heat pumps should occur naturally. However, some natural gas furnaces may continue to be used in cold climates during times when heat pumps exhibit marginal performance.

Additional Environmental Regulations

Anecdotally, chimneys and stacks used to be called smokestacks because what used to come out of them was smoke. Visiting a steel mill overseas in the late 1970s revealed open hearths that were spewing plumes of rust-colored smoke over the town. At the same time, many people in Pittsburgh, Pennsylvania, had a fetish for cleaning their windows because, in the then, not-so-distant past, windows cleaned in the morning became sooty by that afternoon.

More seriously, you may recall that in Chapter 1, a cloud of pollution hung over Donora, Pennsylvania, in 1948, where "nearly half of the town's 14,000 residents experienced severe respiratory or cardiovascular problems. It was difficult to breathe. The death toll rose to nearly 40." [282] It took years, but the response was the passing of legislation to reduce air pollution, which resulted in the installation of pollution control equipment.

At present, the Earth has a global carbon dioxide issue. There appears to be no other viable alternative to generating electricity, other than unpopular nuclear energy, which can produce the amount of reliable near-zero-emission electricity at the scale necessary to meet demand. Unreliable electricity is not the answer, so removing carbon dioxide emissions from existing power generation plants appears to be the only viable alternative with current technology. Everything else appears to be just a bandage that is irritating for the population to put on and keeps falling off to further aggravate the population. Stated differently, positive actions such as turning off the lights when a room is not in use, recycling plastic containers, and eating vegan food will not even make a dent in the magnitude of the global carbon dioxide issue.

Emissions should be treated at their point sources, which are known, typically contain over 10 percent carbon dioxide, and are often already regulated. This approach can mitigate emissions, the monitoring of which can be incorporated into the existing regulatory infrastructure without inconveniencing or depending on the population. The importance of not inconveniencing or depending on the population cannot be overstated because the electric and industrial sectors can handle emission removal without the acrimony, division, infighting, and inconveniences that often characterize change. However, there will be some economic pain.

The regulations suggested above will be unpopular with almost everyone, except those involved with the design, development, sales, and manufacture of the equipment needed for their implementation. For starters, the equipment will be expensive to retrofit into the thousands of existing fossil fuel generating plants, independent power producers, industrial sector, and commercial point sources worldwide. Regulations will also add to the cost of new installations.

Various persuasive arguments against regulation will be suggested. However, similar activities were implemented in the United States in the 1970s, when the legislative response to pollution resulted in the installation of environmental controls such as precipitators, dust collectors, scrubbers, and the like. It took years to see the effects, but the result was clean air and better health.

It is imperative that these additional environmental regulations be implemented in harmony with technological advancement. Implementing the regulations too fast is a recipe for frustration and disaster.

Regulatory carbon dioxide controls will increase the cost of electrical energy supplied somewhat but should not appreciably affect the cost of delivery. Therefore, if the electric supply represented half of a customer's billed amount and electricity supply costs increased by 30 percent due to regulation, the average customer's total electric bill would only increase by 15 percent. The cost of regulatory compliance per industrial product should be much less than the increase in electricity costs because the cost of electricity is usually relatively small when compared to the cost of raw materials and other processing expenditures.

Compliance for fossil fuel electric plants could be achieved using carbon capture, utilization, and permanent storage (CCUS) systems. However, they are not yet fully developed and viable at the scale necessary to satisfy demand. In general, environmental regulations should be implemented in phases over time to soften the economic impact of installation and become comfortable with new technology.

Solar Geoengineering

Solar geoengineering attempts to increase the amount of radiation traveling from Earth to space to cool the Earth. Manifestation can come in many forms that are largely conceptual in nature. Solar geoengineering does not address the issue of increasing carbon dioxide concentration in the atmosphere but rather one of its symptoms, which is global warming.

Examples of solar geoengineering include sulfur dioxide being injected into the upper atmosphere to mimic the effect of a volcano and reflect solar energy back to space, and marine cloud brightening to increase the number of sea salt particles over the oceans to create more white clouds that will reflect more solar energy back to space.

Solar geoengineering research should continue to develop an alternate plan should the concentration of carbon dioxide in the atmosphere or its effects become untenable.

Summary

Carbon dioxide emissions must be lowered to reduce and control the concentration of carbon dioxide in the atmosphere. All the sectors emit carbon dioxide. However, the electric sector and industrial sector contain point sources that should be targeted for mitigation because they emit large amounts of carbon dioxide at relatively high concentrations of approximately 10 percent is easier to remove, as compared to the concentration of almost 1000 times less (420 ppm or 0.042 percent) in the atmosphere. The efficacy of addressing point sources is quantified in the next chapter.

Chapter 18: Potential Emission Reductions

There'll be time enough for countin' when the dealing's done.
The Gambler (song by Kenny Rogers)

Teaching provides an opportunity to share knowledge with others and learn new things, some of which are technical and others personal. One evening over dinner, an instructor told me about an interesting application for a relatively primitive steam valve that he used to control the temperature of liquid leaving a one-meter-long steam heater. [283]

Almost a year after returning to my plant, I was asked to investigate a boiler fuel oil heat set that was not maintaining a steady temperature. One look at the application convinced me to try out the idea, so a steam valve was purchased, installed, put into operation, and received rave reviews for how well it maintained the fuel oil temperature.

My boss later asked me to investigate a poorly controlled steam reboiler that was used to heat a critical 10-meter-tall distillation column that had to operate, or production would stop. A reboiler is also a heat exchanger, so I decided to use the same type of valve used previously on the oil heat set.

The first attempt at using this valve failed because a small internal part of the valve was missing. The second attempt failed because a fitting on the valve leaked. My boss called me into his office and asked if there were real issues to overcome, or if I was just struggling to make something work. I explained the reasons the valve did not operate properly, and he allowed me to install the valve again.

The third attempt worked perfectly. It was so perfect that installing this one valve immediately increased the overall capacity of the plant by 15 percent. Calculations performed after the fact showed that the valve increased profit by approximately one million dollars annually.

In hindsight, it also reduced the carbon footprint of producing the product.

---xxx---

It does not matter whether global warming is occurring or not, or whether it is caused by human activity or not. Denying that the

Earth's atmosphere does not have a carbon dioxide concentration issue is effectively putting one's head in the sand because its concentration is increasing, and in time, the new equilibrium concentration will be dependent on a combination of emissions from natural processes and human activity.

The salient question is whether different human activities can return the Earth to its previous concentration of carbon dioxide. This can be determined by estimating the amount of carbon dioxide emissions that can be mitigated by implementing strategies with the largest potential effect.

Figure 8-1 will be used to analyze each energy source. To simplify the analysis, energy flows of less than approximately 10 percent are ignored. The intent is to identify actions that create a path to return the Earth to its previous carbon dioxide equilibrium with minimal adverse effects on the global population and without impeding advancement to higher standards of living. This transition will not be painless, and ultimate success will require technological advancements occurring over a period of decades.

The changes outlined below are based on the implementation of the strategies in the previous chapter that effectively transform the electric power sector to near-zero-emission processes. They will probably involve some form of carbon capture, utilization, and storage (CCUS) system. The percentages are percentages of the quantities of energy shown in Figure 8-1. No adjustments are made for increases in energy consumption by end-users.

Energy Sector Assumptions

Approximately 75 percent of the petroleum produced is burned, while the remainder is consumed in the industrial sector to make products. If approximately 90 percent of the vehicles are propelled by electricity, the 69 percent of petroleum consumed by the transportation sector would be 7 percent. Approximately 4 percent of a barrel of oil is consumed as aviation fuel, [284] but this will fall somewhat as biofuels become more available. Therefore, the total amount of petroleum burned and producing emissions would fall to approximately 10 percent. Existing petroleum processes would be modified to produce less gasoline and diesel fuel. Zero carbon dioxide emission electricity production would increase by an amount that makes up for the 59 percent drop in petroleum use in the transportation sector.

It is assumed that the residential and commercial sectors will naturally migrate to electric equipment, such as heat pumps, so natural gas consumption in these sectors will fall dramatically. The electricity and industrial sectors will largely operate near-zero-emission combustion processes. Therefore, the amount of natural gas burned, and its emissions are assumed to drop to approximately 10 percent of the amount consumed by the residential and commercial sectors. Near-zero-emission electricity production would increase by an amount that makes up for the 37 percent drop in supply to the electric sector.

It is assumed that the industrial sector will use the same amount of coal. The electricity and industrial sectors will largely operate near-zero-emission combustion processes. Therefore, it is estimated that the amount of coal burned and producing emissions will drop to approximately 10 percent of the amount consumed by the industrial sector. Near-zero-emission electricity production would have to increase by an amount that makes up for the 90 percent drop in supply to the electric sector.

Electric Power Assumptions

Based on the above, the electric power sector will need to upgrade or replace fossil fuel electricity production plants with near-zero-emission combustion processes. This will cause electricity production to be dominated by reliable fuel sources that can ramp up generation when unreliable energy sources are not producing electricity.

It appears that the electric power sector will also need to at least double its current capacity with reliable energy sources to meet the additional demands cited above for the transportation, industrial, residential, and commercial end-use sectors. Somewhat mitigating the need for additional capacity is the ability of electricity generators to not operate as near-zero-emission processes during the relatively few times when the electrical grid operates at or near peak load.

Putting this in perspective, electrification in the 1900s was a process of modifying equipment to operate using electric motors instead of mechanical devices and steam turbines. The evolution of electrification in the mid-2000s appeared to entail changing processes to utilize energy processes that do not emit carbon dioxide.

Potential Carbon Dioxide Reduction

Table 18-1 shows the carbon dioxide emissions in the United States based on Figure 8-1, and the estimated carbon dioxide emissions as if the transition to near-zero-emission electricity was complete.

Table 18-1 United States Carbon Dioxide Emissions Before and After Transition

	Petroleum	Natural Gas	Coal	
Before Transition				
Total Energy	35.1	31.3	10.5	quadrillion BTU
Percent Burned	75	100	100	percent
Annual CO_2 Emissions	1.92	1.66	1.03	billion tons
After Transition (Estimate)				
Percent Burned	10	10	10	percent
Energy Burned	3.5	3.1	1.1	quadrillion BTU
Annual CO_2 Emissions	0.26	0.17	0.11	billion tons
CO_2 Emissions / MMBTU	160 (73)	117 (53)	215(98)	pounds (kg)

The total carbon dioxide emissions before and after the transition of 4.61 and 0.54 billion tons, respectively, represent an 88 percent reduction in emissions. Even burning 50 percent more fuel to remove the carbon dioxide would reduce the emissions before the transition by approximately 82 percent (1.5*0.54/4.61).

223

However, the desired information is global emissions, not just those of the United States. The spaghetti chart for the United States was used to analyze energy processes on a smaller than global scale to enable the identification of the important parameters that should be considered to analyze the potential reduction in global carbon dioxide emissions.

The results are presented in Table 18-2 assuming that the percentages before and after the transition will be similar to those for the United States.

Table 18-2 Global Carbon Dioxide Emissions Before and After Transition

	Petroleum	Natural Gas	Coal	
Before Transition				
Percent Burned	75	100	100	percent
Energy Content[285]	51,170	40,375	44,473	TWh
Energy Content	131.0	137.8	151.8	quadrillion BTU
Annual CO_2 Emissions	9.56	7.30	14.88	billion tons
After Transition (Estimate)				
Percent Burned	10	10	10	percent
Energy Content	13.1	13.8	15.2	quadrillion BTU
Annual CO_2 Emissions	0.96	0.73	1.49	billion tons
CO_2 Emissions / MMBTU	160 (73)	117 (53)	215 (98)	pounds (kg)

The total carbon dioxide emissions before and after the transition of 34.74 and 3.18 billion metric tons, respectively, represent a 91 percent reduction in emissions. Even burning 50 percent more fuel to remove the carbon dioxide would reduce the emissions before the transition by approximately 88 percent (1.5*3.18/34.74). The percentage reduction is different than that of the United States because the global mix of energy sources is different.

The estimated annual carbon dioxide emissions of 3.18 billion tons after the transition is close to the 3 billion tons of carbon dioxide that the Earth could handle until around 1910, per Figure 9-9, so it should be possible to come close to restoring the carbon dioxide in the atmosphere to its pre-1800 concentration.

Reductions of over 80 percent are extreme and will take decades to implement. It will take **considerably longer than 52 years** to remove the accumulated carbon dioxide from the atmosphere. Therefore, returning the atmosphere to its pre-1800 carbon dioxide concentration will take decades longer than it took to clean the air in the United States in the mid-1900s.

Chapter 19: Current Trends

Remember Alice? It's a song about Alice. --- Alice's Restaurant
(song by Arlo Guthrie)

There are some streets that are well lit and others that leave much to be desired. On one such street, only one streetlight was working out of the four or five on the block.

A man on his hands and knees was looking for something under that light. Along came another man, who asked what the first man was doing. The first man looked up and said that he was looking in the grass for a key that he had lost. The second man then asked where he lost the key, to which the first man replied that he lost it over there near one of the non-working lights.

The second man was perplexed and asked why the first man was looking here, when he lost the key over there. The first man replied, "There is light over here."

---XXX---

Consider the following two sentences from the Introduction that represent the majority view on global warming.

Some people claim that burning fossil fuels causes the accumulation of carbon dioxide in the atmosphere, which warms the Earth. This represents an existential climate crisis, so humans should stop burning fossil fuels and immediately pivot to renewable energy sources.

There is sufficient evidence to reasonably conclude that the first sentence is factually correct because burning fossil fuels does warm the Earth and emit carbon dioxide. The second sentence concludes that this is an existential crisis but offers no explanation as to why. The second sentence also states that humans must stop burning fossil fuels immediately, with no explanation.

Now consider the opposing view on global warming from the Introduction.

Others claim that there is no immediate existential climate crisis because carbon dioxide is a trace gas that has not been proven to appreciably affect the temperature of the Earth.

226

This group is correct in that global warming does not pose an immediate existential crisis, but there does appear to be evidence that the concentration of carbon dioxide has appreciably affected the temperature of the Earth. Whether it can be "proven" or not is a matter of how the word "proven" is defined. Proofs in geometry are rigid by design, whereas proofs in conversation can be quite flexible and sometimes prove nothing at all.

What we have is no less than an existential crisis that is not an existential crisis, and global warming that is occurring but is not occurring. The truth appears to lie somewhere in between, where the Earth is warming but will be resilient until a limit is reached, and something happens.

Findings

This book has uncovered the following:

- **Fossil fuels have dramatically increased the standard of living for billions of people** since around 1800. Thousands of products would not exist without fossil fuels.
- Fossil fuels have increased and continue to increase the concentration of carbon dioxide in the atmosphere.
- Global population growth is projected to slow around 2060 and stabilize at approximately 10.5 billion around 2100.
- Global per capita energy consumption stabilized, with decreases in North America and Europe offsetting increases in China and India. This stabilization is expected to be temporary because population growth and rising standards of living, offset somewhat by advances in technology, are expected to increase global energy consumption.
- It took 110 years to accumulate approximately 940 gigatons of carbon dioxide in the atmosphere. With zero carbon dioxide emissions, its **removal will take approximately 100 years**.
- The **concentration of carbon dioxide in the atmosphere will reach 1000 ppm in approximately 251 years** at its rate of accumulation in 2023. Therefore, the increasing concentration of carbon dioxide in the atmosphere per se is **not an immediate existential crisis**. However, it should be addressed in the coming decades by developing near-zero-emission combustion processes. Other remedies are helpful

and should be pursued. However, collectively, they have a relatively small impact on emissions.

- When near-zero-emission combustion processes become viable, legislation to control carbon dioxide emission sources can drastically reduce emissions without requiring cooperation from the population.
- **Fossil fuels will be consumed in large quantities for decades to come.**
- Satellite measurements indicate that significantly less global warming may have occurred as compared to that indicated by surface measurements.
- Hybrid vehicles in the mid-2020s will have a smaller lifecycle carbon footprint than gasoline vehicles, a similar or smaller lifecycle carbon footprint as compared to electric vehicles, a smaller mineral content than electric vehicles, and be less expensive than electric vehicles.

These findings should be seriously considered before deciding which actions are appropriate to mitigate the real problem at hand.

Semantics can play an important role in shaping actions to address climate change. For example, there have always been extreme weather events, but *extreme weather events* are now almost exclusively associated with climate change and global warming. *Zero-emission vehicles* are now assumed to have no carbon footprint, but they do have one. *Electric vehicle batteries* are seemingly environmentally clean devices, but they are not.

Above all, it would be prudent to listen carefully to what people say, and how they say it. Do not get carried away by Rosling's *fear instinct* or *urgency instinct*. **Above all, think for yourself before accepting the conclusions of others**.

Malthusian versus Cornucopian Worldviews

Over two centuries have passed since Malthus wrote about the population not being able to grow beyond what its resources can support. Stated differently, Malthus viewed people as more mouths to feed in a world where the food supply is limited. Sustainable development is derived from the Malthusian worldview desire to limit the use of fossil fuels, promote solar and wind energy, and advocate for population reduction, all under the auspices of a centralized controlling structure. These thoughts tend to dominate many institutions, including schools, academia, government, some businesses, and most of the media, in the mid-2020s.

In stark contrast, the Cornucopian worldview holds that technological progress is the key to meeting the needs of a growing population. People are viewed as more brains to innovate and more hands to develop economic growth. Higher prices for scarce resources incentivize conservation, while new products and processes use scarce resources more efficiently or eliminate the need for them altogether.

The organization and content of this book somewhat parallel the evolution of global warming and climate change. Chapters 3 to 5 describe life in the world of Malthus. Chapters 6 to 10 cover technical developments resulting from a Cornucopian worldview. Climategate and other conflicts between these worldviews appear in Chapter 11. Subsequent chapters develop a Cornucopian approach to resolve the issue, with the understanding that Malthusian limits do exist and may have to be addressed.

The debate over global warming and climate change can be viewed as another front in the battle between the Malthusian and Cornucopian worldviews. Decades-old Malthusian predictions, such as running out of oil, have not materialized, while Cornucopian predictions of improving the human condition and growing global wealth have prevailed. Both Malthusian and Cornucopian worldviews exhibit Rosling's *single instinct*, where the same tool is used for every task.

Despite what appear to be clearer perceptions of the world, more accurate predictions, and superior solutions, as of the mid-2020s, proponents of Cornucopian worldview have lost control of most institutions and many businesses to those tending to support Malthusian worldview.

Emission Trends

Since the 1970s, progress has been made in many countries to reduce their per capita carbon dioxide emissions, even though their populations have grown significantly.

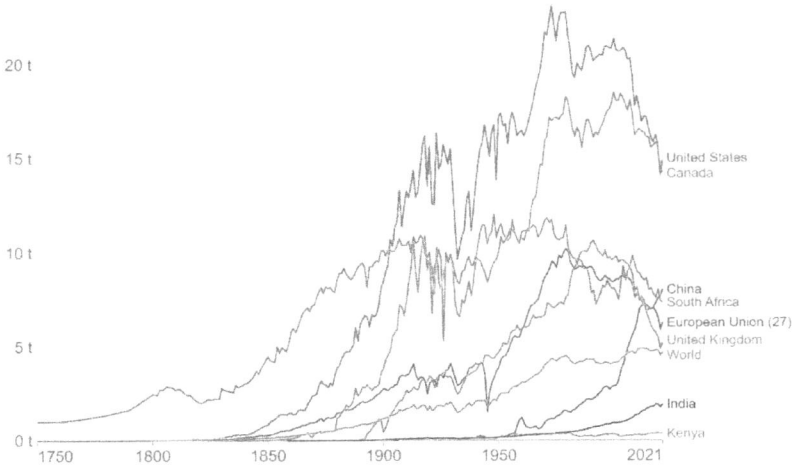

Per capita CO₂ emissions

Carbon dioxide (CO₂) emissions from fossil fuels and industry¹. Land use change is not included.

Figure 19-1 Per Capita Carbon Dioxide Emissions [286]

Figure 19-1 shows that global per capita carbon dioxide emissions may have stabilized around 2010. India's per capita emissions have increased. However, per capita emissions in China have increased significantly, and appear to be rising. Global per capita emissions appear to have stabilized since around 2000.

Figure 19-2 shows that global carbon dioxide emissions appear to be increasing. In 2021, China, India, and the remainder of Asia represented approximately 31, 7, and 20 percent of global carbon dioxide emissions, respectively, totaling almost 60 percent of global emissions. The populations of China and India together comprise approximately 36 percent of the world's population [287] and produce approximately 38 percent of the world's emissions. There is only so much that the United States can do alone when it emits less than 14 percent of global emissions, as per Figure 19-2.

China and India were not included in the Kyoto Protocol list of countries with limits and reductions on greenhouse gases. Therefore, despite producing almost 40 percent of all global emissions, Figure 19-2 shows that China is the world's largest emitter but remains exempt from any pollution limits.

230

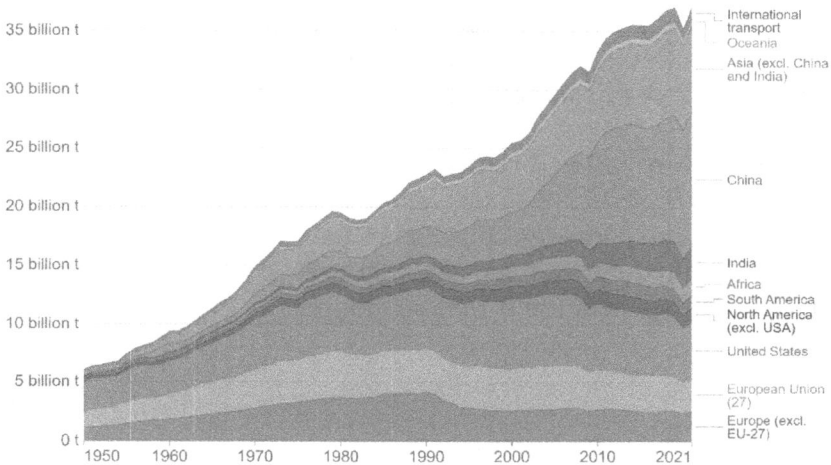

Annual CO₂ emissions by world region
This measures fossil fuel and industry emissions¹. Land change is not included.

Source: Our World in Data based on the Global Carbon Project (2022)   OurWorldInData.org/co2-and-greenhouse-gas-emissions • CC BY

1. **Fossil emissions**: Fossil emissions measure the quantity of carbon dioxide (CO₂) emitted from the burning of fossil fuels, and directly from industrial processes such as cement and steel production. Fossil CO₂ includes emissions from coal, oil, gas, flaring, cement, steel, and other industrial processes. Fossil emissions do not include land use change, deforestation, soils, or vegetation.

Figure 19-2 Annual Carbon Dioxide Emissions by World Region [288]

Both China and India need abundant, scalable, reliable, and flexible energy that is affordable for their people as they improve their standard of living. Both may say that they will be cognizant of the environment, but both have vast coal reserves. Adding significant amounts of coal-burning electric generation capacity will emit tons and tons of carbon dioxide during their operating lives. [289]

There are some who opine that China and India will never move away from fossil fuels this century unless a better alternative is created. Renewables are only meant to placate gullible western environmentalists, foundations, academia, and governments. [290] This has yet to be proven.

Predicting emissions when the global population stabilizes around 2100 is fraught with unknowns. Per Figure 19-1, the global per capita carbon dioxide emissions were less than 5 tons in 2021. The average per capita emissions of Canada, the United States, the European Union, and the United Kingdom was approximately 10 tons in 2021. Per capita emissions in India are approximately 2 tons in 2021, but they are expected to grow as living standards increase.

China's per capita emissions are approximately 8 tons and growing. Per capita emissions in Africa are low but are expected to grow substantially as its population grows. Left unchecked, it would not be surprising if global per capita carbon dioxide emissions doubled or tripled by 2100. Innovation and mitigation during the next 75 years will reduce this increase, and perhaps reduce emissions.

Electricity Generation Trends

Figure 19-3 shows that the amount of electricity derived from solar, wind, and bioenergy sources is small but increasing, while oil, nuclear, and hydropower remain relatively stable. Natural gas and coal electricity generation is trending higher.

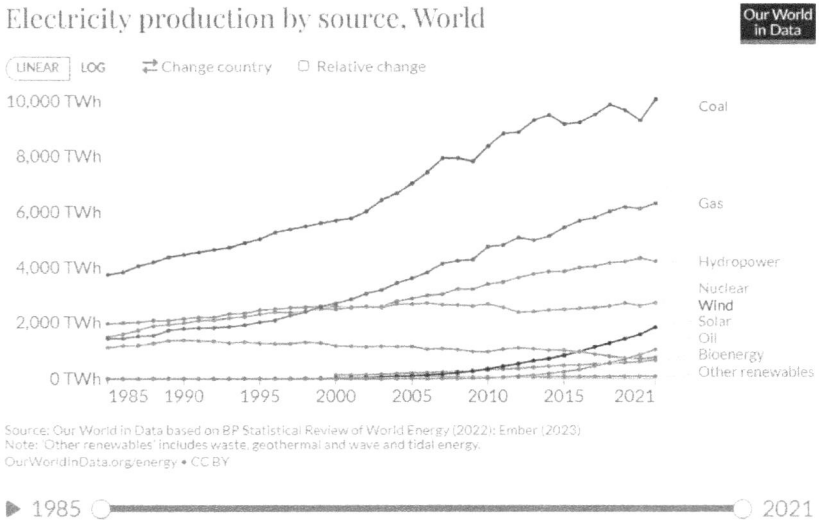

Figure 19-3 Electricity Production by Energy Source [291]

Global electricity production was 27,813 TWh in 2021. Coal and natural gas generated 10,086 and 6338 TWh, which accounts for the generation of 36 and 23 percent of global electricity, respectively. Both will likely continue to trend higher due to increasing coal generating capacity in China and India, and the high efficiency associated with combined-cycle natural gas electricity generation.

Deaths From Natural Disasters

Figure 19-4 shows that per capita deaths from natural disasters have generally decreased since 1900. Materials made directly from fossil fuels and knowledge developed indirectly from consuming fossil fuels enabled much of this reduction.

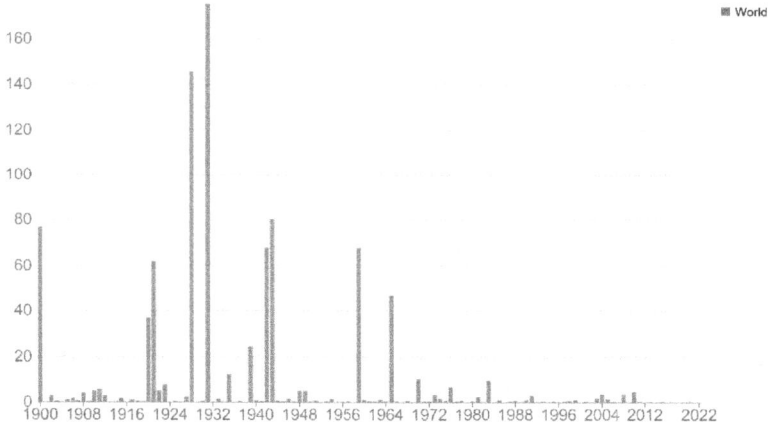

Death rate from disasters

Death rates are measured as the number of deaths per 100,000. Disasters include all geophysical, meteorological and climate events including earthquakes, volcanic activity, landslides, drought, wildfires, storms, and flooding.

■ World

Figure 19-4 Per Capita Death Rate from Natural Disasters [292]

Paths Forward

One can debate whether to take certain actions or not. However, what is not debatable is that moving too fast will cause harm to the population, if the technology involved is not fully developed, or when a new energy source and its infrastructure are not in place. Incorrect and inappropriate actions will result in worse consequences. **Of particular importance is not to legislate too quickly**, because creating emission standards that are either too expensive or difficult to meet will be expensive and stifle progress.

Headwinds to reducing carbon dioxide emissions include a population that continues to grow with rising standards of living, which will increase energy demand, especially in China and India, which are constructing new coal-burning power plants. It is expected that carbon capture, utilization, and storage (CCUS) will play a significant role in reducing carbon dioxide emissions after

233

additional research and development improve efficiency and economic viability.

Tailwinds include population stabilization around 2100, more efficient utilization of energy and resources, and Level 4 countries reducing their carbon footprint. Of perhaps greater importance is the understanding that we do not know whether existing technologies will become viable or whether innovators will discover entirely new technologies and processes that mitigate the accumulation of carbon dioxide in the atmosphere similar to how the automobile mitigated the Great Horse Manure Crisis of 1894 in Chapter 3.

Benjamin Franklin famously tied a key to a kite in the mid-1700s, and Alessandro Volta invented the battery around 1800. The electric motor and light bulb were invented in the early and late 1800s, respectively. The first nuclear power plant was opened in the 1950s. In the early 2000s, researchers were investigating algae that produce electricity. There have been thousands of important developments in between.

Technology has advanced exponentially since 1800 and should continue to do so into the future. The Cornucopian key to addressing the concentration of carbon dioxide in the atmosphere can be summed up in one word: technology. The caveat is that there may be Malthusian limits that need to be addressed with Malthusian solutions along the way.

Chapter 20: Climate Change Politics

Never let a good crisis go to waste. --- Winston Churchill

The project cost overrun described in Chapter 15 was too large to hide during the final project review process. Someone or some group had to be blamed for the overrun. The civil, structural, mechanical, piping, and electrical groups had all far exceeded their budgets because there were many more foundations, buildings, equipment, pipes, and motors that were added to the original design. However, the instrumentation portion of the project came in near the original budgeted amount. How could that happen? After all, the additional equipment required the purchase of additional instruments.

My boss estimated the original project using conventional controls, as was common practice at the time. At the time the project was implemented, computer-based controls had become more accepted and affordable. A major difference between the two controls was that adding another conventional controller involved purchasing and physically installing another controller. Implementing the same functionality in computer-based controls amounted to adding the controller in software and on a display, effectively at no cost. Therefore, the cost of the additional instruments on this project was offset by the lower cost and expandability of the computer-based controls.

My boss was blamed for the overrun, along with the managers of the other groups. In addition, he was ostracized by the other managers, because he did not overrun his budget. After all, he made them look bad.

---xxx---

There is no rule stating that policy decisions must be made based on scientific findings. History is littered with policies that were just plain wrong, just as many policies relating to global warming and climate change are wrong, address a different problem, or benefit decision makers and their friends.

For example, California should have installed nuclear electric generation plants years ago, but no American politician has killed more clean energy than the governor of California - and in ways that often benefited his own family financially. [293] California's electricity prices are among the highest in the United States.

Another example is that hybrid vehicles should be the vehicle of choice in the mid-2020s, as per Chapter 14. However, electric vehicles have been heavily promoted and subsidized, while the more flexible, less expensive hybrids that contain less mineral content and can have carbon footprints similar to electric vehicles are often not mentioned.

Bad policy decisions typically waste money and make things worse, often in insidious ways that, unbeknownst to the population, increase prices and disproportionately affect people living below Level 4. Stated differently, both good and bad policy decisions determine the outcome for the population and the environment.

Loss of Personal Freedoms

The Limits to Growth (1972) was the clarion call for activism to better understand and address global economic, political, natural, and social issues. Since its publication, people associated with The Club of Rome have been spreading awareness of these issues and promoting its tenets. Repeating from Chapter 11, The Limits to Growth (1972) concluded that (**emphasis added**):

1. If the present growth trends in world population, industrialization, pollution, food production, and resource depletion continue unchanged, the limits to growth on this planet will be reached sometime within the next one hundred years. **The most probable result will be a rather sudden and uncontrollable decline in both population and industrial capacity**.
2. It is possible to alter these growth trends and to establish a condition of ecological and economic stability that is sustainable far into the future. The state of global equilibrium could be designed so that the **basic material needs** of each person on earth are satisfied and each person has an **equal opportunity** to realize his individual human potential.
3. If the world's people decide to strive for this second outcome rather than the first, the sooner they begin working to attain it, the greater will be their chances of success. [294]

The first conclusion above mirrors that of the Malthusian worldview, discussed previously in Chapter 4, where Malthus wrote that "an obvious truth (is that the) ... **population must always be kept down to the level of the means of subsistence.**" [295] Interestingly, the second conclusion above suggests an equilibrium satisfying the *basic material needs* of food, water, clothing, and shelter. By observation, people living at Rosling's Level 1 have their basic material needs satisfied, or they would be dead. Similarly, people living at Level 1 have equal opportunity and potential, albeit they are overwhelmingly poor.

From Chapter 11, salient preliminary views (**emphasis added**) expressed in The Limits to Growth (1972) include that **mankind should seek a state of equilibrium**, technological solutions alone cannot be expected to suffice, a **rapid, radical redressment** of an unbalanced and dangerously **deteriorating world situation is the primary task** facing humanity, effort must be resolutely **undertaken without delay,** the major **responsibility must rest with the more developed nations**, and the state of equilibrium must ultimately be founded on a **basic change of values and goals at individual, national, and world levels.** [296]

By observation, these conclusions could form the basis of recommendations to **reduce the population and use less resources,** such as effectively **moving people from Level 2, Level 3, and Level 4 back to Level 1,** where everyone will live at a subsistence level with basic needs satisfied and equal opportunity. The Limits to Growth: The 30-Year Update (2004) reinforces this notion. From Chapter 11 (**emphasis added**), if the industrial world begins to act upon two definitions of *enough*, one having to do with material consumption, the other with family size, the **world population can achieve a level of well-being roughly equivalent to the lower-income nations of present-day Europe,** [297] **whose median income is approximately 30 to 40 percent that of the United States.** [298]

This approach has headwinds and tailwinds. The approximately 1 billion people living at Level 1 would benefit from a uniform standard of living, while the 3 billion at Level 2 might not be affected much. However, **what is not said is often more important than what is said.** Most of the 2 billion at Level 3, and especially the 1 billion at Level 4, would be devastated as their standard of living plummets and they see their hard work and sacrifices beget the same standard of living as everyone else, many of whom produce little. Worse yet, not benefiting from the fruits of their labor will negatively impact their drive to succeed. Worse yet, creativity and innovation will be stifled. Worse yet, the difficulties of their lower standard of living will create and exacerbate serious economic, emotional, and safety problems at home, and negatively affect families.

Implementing a uniform standard of living, even with some differences between countries, would effectively tear down much of our existing society, only to start anew. The downward movement of living standards will be strongly resisted. By way of example, people in the United States will have to reduce their income by approximately 60 to 70 percent to reach parity with the lower-income European countries. In contrast, reaching parity with the lower-income European countries would be a huge step up for the average person living in many African, Asian, and South American countries.

Repeating from Chapter 11, The Club of Rome stated (**emphasis added**):

> **Now is the time to draw up a master plan for organic sustainable growth and world development based on global allocation of all finite resources and a new global economic system.** [299]
>
> Increasing independence between states and regions must then translate as a decrease in independence. **Nations cannot be interdependent without each of them giving up some of, or at least acknowledging limits to, its own independence.** [300]

Implementation will require a system in which private property and the distribution of income are subject to social control. This is the definition of *socialism* [301] and will likely not be compatible with the United States Constitution or laws in other countries.

Repeating from Chapter 11, the 2022 update of Limits to Growth suggests adopting *Solidarity Capitalism*. (**Emphasis added**)

Solidarity Capitalism as an idea, or academic concept, would encourage autonomous thinking by **aligning the sense of community and social responsibility with the capitalist model**. [302] Solidarity Capitalism as an ethical concept, as well as an economic model, ensures that we accompany the **switch from a shareholder-centric to a stakeholder-focused model** with ethical principles that include fairness and equity. [303]

To remodel modern capitalism as a system that serves humanity and the planet... [304]

Capitalism is an economic system where industry is owned privately and operated for profit. Although not certain, the Solidarity Capitalism concept might allow private ownership, but operations will be based on the betterment of humanity. This is a disincentive for owners, who will receive smaller profits when focusing on stakeholders, instead of shareholder profit. In many cases, an otherwise profitable business can lose large amounts of money, so the owner may have to sell his newly downgraded house to cover the loss.

For example, the commissioner of the Consumer Product Safety Commission said that natural gas stoves are a hidden hazard, and products that can't be made safe can be banned. [305] Where does that leave millions of people who want consumer choice and prefer gas stoves? What happens when their energy costs increase when they are forced to use another technology? Will the factories that make gas stoves be driven out of business? Approximately 76 percent of methane emissions from natural gas stoves occur when the stove is off. [306] If this is the overriding issue, perhaps better-designed shutoff valves would mitigate most emissions at minimal additional cost without disrupting the population.

Unobtrusive language such as 'seeking a state of equilibrium', 'resolutely undertaken without delay', 'dangerously deteriorating world situation', and 'basic change of values and goals', are nice ways of telling you why **your wealth** is being confiscated and your income is controlled by others.

Wealth Transfer

The previous section suggests that humans are on a collision course with disaster because current human consumption is not sustainable. Fewer people create less pollution, so one remedy is to reduce the population. Wars, disease, and famine kept the population sustainable for thousands of years. More modern approaches include family planning, access to birth control, adoption instead of having children, having children later in life, abortion, sterilization, assisted suicide, and fostering lifestyles that do not produce offspring. Another remedy is a readjustment that imposes the same standard of living on everyone to make human consumption sustainable, where newly poorer people pollute a lot less and newly better-off people improve their lot without emitting much additional pollution.

Placing the burden of redress on a certain group of countries suggests a *transfer of wealth* from a group of countries that are well-off to another group that is less fortunate. Why is the second group less fortunate? Perhaps a hurricane devastated the country, but it is more likely the result of poor governance, corruption, and a lack of incentive that caused a lack of opportunity and good-paying jobs, which are often reasons why people in some countries do not do well.

Repeating from Chapter 4, the following are modified versions of the five fundamental activities that immigrant groups in the United States and people in general should embrace to achieve a higher standard of living, understanding that people in other countries should learn their own language. [307]

- Be honest and obey the law.
- Work hard.
- Be self-reliant and do not accept government assistance.
- Learn to read, write, and speak excellent English; learn the culture well.
- Obtain the most advanced education possible in a field that is in demand for employment.

It might prove helpful to ask if a country and its people embrace these activities. The first four do not require much money to implement but rather a basic infrastructure and the determination to improve one's standard of living. Attending a university could be beyond the means of some, but enrolling in night classes at a local college is often possible. Nonetheless, **educating the population is key** to improving standards of living. Handouts, such as wealth transfers, are one-time windfalls that may help temporarily but not have a lasting effect.

For example, around World War II, Singapore was a remote mosquito-infected British outpost on a small island, and Palestine (previously part of the Ottoman Empire and now Israel) was a backward British colony that was largely a desert. Both countries were small with populations of approximately 1 million people [308] [309] and no appreciable natural resources. However, both developed into vibrant countries in a relatively short period of time, in large part due to an educated population and the rule of law. In contrast, many African countries have immense natural resources but are not doing well. *Resources are always limited, even for advanced countries, but the problem in Africa is not resources per se. The problem is their misuse.* [310]

Individuals who do not implement all five tend to not do well, even in countries that are well off. A transfer of wealth may bring some benefit, but it does not teach people how to sustain themselves in the long term and more importantly, influence future generations to do the same. As Chinese philosopher Lao Tzu, founder of Taoism, said, "Give a man a fish and you feed him for a day. Teach him how to fish and you feed him for a lifetime". [311]

If the population of a country is not fostering all five of the above activities, what does transferring wealth to these countries hope to achieve? This begs the question of the wisdom of transferring wealth to countries that are not prepared to use it effectively.

Malthusian versus Cornucopian Worldviews (Revisited)

Malthusians view people as mouths to feed in a world where the food supply is limited. Cornucopians view people as more brains to innovate and hands to produce. Unfortunately, groups that espouse one worldview develop solutions that do not incorporate solutions developed from the other worldview. The best solutions often contain elements of both.

Neither worldview is always correct or always wrong. Malthusians identified limits that led to solutions that mitigated air and water pollution in the mid-1900s and put the depletion of the ozone layer on a path to recovery. These problems were solved in different ways. Pollution was a local problem that was addressed by the installation of pollution abatement equipment on a global scale, a solution that combines Malthusian limits with a Cornucopian solution. In contrast, the depletion of the ozone layer was addressed using a Malthusian solution by eliminating the manufacture of certain chemicals because no Cornucopian solution was available.

On the other hand, most potential limits were addressed using Cornucopian solutions. Therefore, it is advisable to first consider Cornucopian solutions with the understanding that definitive Malthusian limits can sometimes occur and may need to be addressed.

Chapter 11 suggests that global warming may or may not exist, and if it does exist, it may be considerably smaller than previously thought. The concentration of carbon dioxide in the atmosphere may or may not be within a normal range, but if normal, it will soon exceed its normal limit if it maintains its current trajectory. Given the positive past performance of Cornucopian solutions and the anticipated stabilization of the global population, it would be prudent to apply Cornucopian solutions, such as those presented in this book, to control the carbon dioxide concentration in the atmosphere before resorting to disruptive Malthusian actions, such as restricting the use of carbon fuels, reducing living standards in much of the world, and taking freedom away from many people. However, Malthusian actions might eventually become necessary if a viable Cornucopian solution is not identified in time.

Origin of Tactics

Some tactics employed by global warming and climate change advocates can be traced back to slavery, whose history spans many cultures, nationalities, and religions from ancient times to the present day. Likewise, its victims have come from many different ethnicities and religious groups. The social, economic, and legal positions of enslaved people have differed vastly in different systems of slavery at different times and in different places. [312] However, a common thread is that **dependence on their master was fostered, in part, by limiting education and information.**

Regardless of the subject, spreading an agenda thrives on a combination of limited critical thinking skills and a lack of information. Knowing when to investigate further and how to analyze additional information can validate or debunk many claims. Other claims may require further independent investigation and research.

A request for data could easily debunk the example in Chapter 2, where in 2019, a sitting member of the United States Congress said that the world is going to end in 12 years if we don't address climate change... [313] Calculations in Chapter 10 show otherwise but, asking for data and conferring with levelheaded scientists before writing a news article would seem appropriate and provide balance. In 2018, at the age of 15, child activist Greta Thunberg promoted a scientist's warning that "climate change will wipe out all of humanity unless we stop using fossil fuels over the next five years." [314] Over five years have passed, and humanity is still around.

The need for additional investigation is illustrated by the initial interpretation of Figure 4-1, which suggested exponential population growth. A more detailed examination revealed that the growth rate was falling, contradictory conclusions resulted from heat wave data in Figure 10-8 and Figure 10-9, and the existence of credible challenges to the data used to show the existence of global warming that were presented in Chapter 11. These examples are usually not identified by those reporting about global warming but rather by a small number of people who raise doubt and investigate when something does not seem to be correct.

A large segment of the population has limited exposure to science and may not know that virtually all products in common use have been made from, or with, fossil fuels. Most people would not know what questions to ask in a discussion about global warming, and their opinion is often based on what someone tells them, and not necessarily on logical thought. This reality is being used to foster dependence on politicians, reporters, and so-called experts, who often receive funding to research global warming and climate change. This dependence is analogous to a country, such as North Korea, which withholds information that it does not want its people to see. To wit, global warming or climate change is raised in almost every story about weather events *considered* extreme, where almost none of them would be considered extreme if someone took the time to evaluate them in context.

Irrelevant Facts

Chapter 11 addresses researchers and scientists who question the extent to which global warming is occurring and how it is measured. For the most part, they present reasonable questions based on data and observations that can be rebutted scientifically, or not. However, many people and organizations have opinions and state facts, but rarely bring data to back them up or explain why they are relevant, respectively.

For example, some might argue that carbon dioxide is just a trace gas in the atmosphere. This is a true statement, but the implication is that the carbon dioxide concentration is so small that there is effectively no effect on the Earth. However, energy from the Sun passing through a vertical column containing a concentration of 420 ppm is equivalent to passing through more than one meter of pure gas. This is not necessarily insignificant.

They may say that global warming happened naturally a few times and that the Earth was much warmer long before humans burned fossil fuels. This is true, but the Earth does not have the same flora and fauna as it did tens and hundreds of millions of years ago, when no humans were living.

They might say that global warming is not an immediate crisis. Correct again. However, the growing concentration of carbon dioxide in the atmosphere will become a problem as it rises closer and closer to compromising human health. Therefore, humans should take action to mitigate it when it is practical and while there is time to do so. We did this locally with air and water pollution in the mid-1900s and, more recently, globally with the ozone layer. [315]

The claims seem to never stop. However, the information in this book provides a prism through which most of the claims can be understood and considered fairly.

Additional literature questioning the existence and relevance of global warming can be found in this endnote. [316]

Government Meddling

Mandating, legislating, and regulating specific behaviors or purchases, whether outright or indirectly, can be helpful but usually are counterproductive. Mistakes made in markets tend to be small and self-correcting. Mistakes made by governments tend to be large and more likely to have catastrophic effects. [317] Government bureaucracies predictably fall victim to regulatory capture, tunnel vision, moral hazard, and corruption. [318] Stated differently, the government can overreach in scope, overstep its authority, and cause damaging results.

Environmental regulations can be manipulated for purposes beyond protecting the environment and effectively mandate behaviors. For example, past increases in vehicle gasoline mileage requirements allowed vehicle manufacturers time to incorporate energy-efficient technology into their vehicles. Excessively large increases provide an unnecessary headwind for vehicle manufacturers.

Similarly, mandating electric vehicles may seem appropriate, but it can adversely affect the environment and be a hardship for most of the global population who live at Level 2 and Level 3, who have a tough time making ends meet. It would not be pleasant for people living at Level 4, either. Many people would not support mandating electric vehicles if they understood the practical and environmental effects of an accelerated universal adoption.

Mandates can be used to achieve political objectives that have little to do with science, despite claims to the contrary. An Executive Order was issued by the President of the United States on 20 January 2021, that reads, in part (**emphasis added**):

> Section 1. Policy. Our Nation has an abiding commitment to empower our workers and communities; promote and protect our public health and the environment; and conserve our national treasures and monuments, places that secure our national memory. Where the Federal Government has failed to meet that commitment in the past, it **must advance environmental justice**. In carrying out this charge, the Federal Government must be **guided by the best science** and be protected by processes that ensure the integrity of Federal decision-making. It is, therefore, the policy of my Administration to listen to the science … and **to prioritize both environmental justice and the creation of the well-paying union jobs** necessary to deliver on these goals.[319]

For reference, the United States Department of Energy defines environmental justice as the fair treatment and meaningful

245

involvement of all people, regardless of race, color, national origin, or income, with respect to the development, implementation, and enforcement of environmental laws, regulations, and policies. [320]

The Executive Order goes on to revoke the permit for the nearly complete Keystone XL pipeline, so crude oil that would have traveled from Canada to the United States through the pipeline, will continue to be transported on less efficient and more dangerous trains. The reasons for the cancellation include undermining U.S. climate leadership by undercutting the credibility and influence of the United States in urging other countries to take ambitious climate action, accelerating the transition toward a clean energy economy, and not being consistent with the Administration's economic and climate imperatives. [321] Scientifically speaking, none of the cited reasons are based on science. However, the objectives appear to have been met via arguments that are eerily similar to the circular argument presented at the beginning of Chapter 2, where the desired conclusion is stated as fact and then reasoning that leads to the desired conclusion is presented. Various other governmental agencies and groups have tried to halt construction or shut down other pipelines using similar arguments.

Mandating change has already been tried and failed miserably. California's mandate to require automakers to provide electric vehicles was **repealed in 1996 (emphasis added)**, [322] ostensibly because the technology was not sufficiently mature for production cars. Thirty years later, the technology is available, but electric vehicles are expensive, potentially emit more carbon dioxide than a hybrid vehicle, as suggested in Chapter 14, and, by at least one account, may also cost more to operate than a comparable car that runs on gasoline. [323]

California is at it again by mandating that 35 percent of new cars sold in California be electric vehicles in 2026, escalating to 100 percent in 2035. [324] In addition, by executive order, half of the medium-duty and heavy-duty trucks sold in California will be zero-emission in 2035, escalating to 100 percent in 2045, [325] [326] which will effectively eliminate the use of gasoline and diesel fuel in the state.

California touts these vehicles as having zero emissions. This is true, but what these vehicles really have is zero tailpipe emissions. The electricity used to operate them generates emissions. But wait a minute. California had approximately 81.7 GW of generating capacity in 2021 [327] and plans to shut down 3.7 GW of natural gas generation and 2.2 GW of its last nuclear power plant by 2030. [328] It also appears that no new fossil-fueled power plants will be built. [329]

This begs several questions, such as how California will provide electricity to ten times more electric vehicles in addition to a fleet of approximately 1 million medium-duty and heavy-duty electric trucks [330] when it is already asking electric vehicle owners not to recharge their vehicles during peak hours. [331]

Electric trucks typically travel 200 to 300 miles on a charge and can take 90 minutes to re-charge. [332] New designs can travel 300 to 500 miles on a charge, and then re-charge to 70% in 30 minutes. [333] In contrast, trucks can travel approximately 900 to 1200 miles on a full tank of diesel fuel. [334] How much productivity will be lost due to the time required to recharge truck batteries? Where will the trucks recharge outside of California, where charging stations will generally be fewer in number? Will California have sufficient generating capacity to power additional electric stoves, electric water heaters, and electric heat pump systems?

The short-term answer appears to be to import electricity from other states. This strategy would effectively allow California to maintain its low in-state emissions by exporting emissions for its electricity to other states that burn natural gas and coal. In addition, electric vehicle batteries, solar cells, and wind turbines are not manufactured in California, so their relatively large carbon footprints are also exported out of state. California might achieve net zero emissions in-state. However, claiming such net zero emissions would be disingenuous if California effectively exports its emissions to other states and countries.

By observation, requests to not charge electric vehicles and delaying the closure of its nuclear power plant are indications that the process of importing electricity is not working out well. California does plan to install renewable electricity capacity to mitigate the situation. However, overdependence on unreliable wind and solar electricity will likely prove to be inconvenient, expensive, and potentially catastrophic.

Maybe these policies will work and maybe not, but it is likely that many people and businesses will purchase their vehicles in other states, even as the availability of fossil fuels in California dwindles with the number of fossil fuel vehicles. In addition, expensive, unreliable electricity might also encourage people to leave California.

Not to be outdone, New York State was the first state to ban natural gas appliance hookups. As previously discussed, these appliances would be phased out naturally as the availability of near-zero-emission electricity increases and (indirectly) as heat pump technology evolves. Sometimes, the better course of action is to avoid conflict and disruptions by doing nothing and letting natural forces play out, even if it takes a little longer.

Even the military suggested considering that "heavy battle tanks or fighter jets and naval ships ... consume a lot of fossil fuel ... and therefore we do have to look into how we can reduce those emissions by alternative fuels, solar panels, other ways of running our missions." [335] Designing mission-critical equipment to reduce emissions creates a myriad of logistical obstacles. For starters, where are the charging stations? How long will a recharge take? What if the solar cells get dirty? Can solar cells survive extreme vibration? What are the effects of bombs exploding nearby?

Not to be outdone, "Britain is developing electric combat vehicles. Not just because they're better for the environment than those old gas-guzzling, carbon-emitting tanks. But also, because it will make the military a more attractive career for a new generation of prospective recruits who are passionate about issues such as climate change." [336] There is a general understanding that the military should focus on combat missions and not the *satisfaction of their recruits*.

The following is an excerpt from the New York City statute that would enable the regulation of emissions from approximately 100 coal-fired and wood-burning pizza ovens, mentioned previously in Chapter 11. (**Emphasis added**)

> §24-105 General powers of the commissioner. (a) Subject to the provisions of this code, **the commissioner may take such action as may be necessary** to control the emission of any air contaminant [which] that causes or may cause, by itself or in combination with other air [contaminant] contaminants, detriment to the safety, health, welfare or comfort of the public or to a part thereof, injury to plant and

animal life, or damage to property or business. The commissioner may exercise or delegate any of the functions, powers and duties vested in him or her or in the department by this code. **The commissioner may adopt such rules, regulations and procedures as may be necessary** to effectuate the purposes of this chapter, including rules, regulations and procedures to establish fees and to authorize and **encourage the development and use of environmentally beneficial technologies.** [337]

It appears that the commissioner can unilaterally do virtually anything to address emissions and even advocate for certain technologies.

The War on Fossil Fuels

The following puts fossil fuels into perspective:

> As the world has grown more prosperous, threats to human security have become less common. The prosperity that fossil fuels make possible, including helping produce sufficient food for a growing global population, is a major reason the world is safer than ever before. [338]

Many groups, ranging from non-governmental organizations (NGO) to political parties to government officials, are proponents of utilizing only renewable electricity sources such as geothermal, hydropower, solar, waste-to-energy, wave, and wind energy, to the exclusion of fossil and nuclear fuel sources. As a factual matter, geothermal and hydropower are not available in sufficient quantities to meet demand, and the remaining renewable energy sources are unreliable or not fully developed.

Nonetheless, eleven (11) United States Senators introduced a resolution:

> Recognizing the duty of the Federal Government to create a Green New Deal by ... meeting 100 percent of the power demand in the United States through clean, renewable, and zero-emission energy sources, including -- (i) by

249

dramatically expanding and upgrading renewable power sources; and (ii) by deploying new capacity. [339]

This is not possible with current technology and may never be possible. Despite data to the contrary presented in Chapter 10, the same resolution also says that "a changing climate is causing ... an increase in wildfires, severe storms, droughts, and other extreme weather events." [340]

This resolution promotes stopping the burning of fossil fuels specifically for generating electricity. Other groups are even more extreme and want to eliminate fossil fuels altogether. This may sound great on paper, but it is ridiculous not only on its face, but on your face as well. Your glasses, contact lenses, goggles, cosmetics, lipstick, face moisturizer, medicinal cream, suntan lotion, wrinkle remover, toothpaste, toothbrush, fillings, braces, and hearing aids would largely not exist, if not for fossil fuels. Just about everything around us is derived from or uses a derivative of fossil fuel in its production. A total ban on coal, oil, and gas would eliminate the existence of thousands of everyday products, put the world on a fast track toward Level 1, and significantly lower everyone's living standards.

People, groups, and platforms that promote banning fossil fuels, or energy sources that are not viable, cannot be scaled to meet demand, or are unreliable for more than supplementing reliable electricity, are not knowledgeable, disingenuous, or both. People developing such platforms should know better.

It appears that just about anything can be affected by climate change and the attacks can come from seemingly anywhere. For example, it was reported that the Attorneys General of Kansas, Oklahoma, and Texas filed a lawsuit in federal court because of the:

> ... listing of the lesser prairie chicken as a threatened or endangered species by the Biden administration's U.S. Fish and Wildlife Services (USFWS). The listing covers the entirety of the bird's habitat—which includes the southwest quarter of Kansas as well as the panhandles of Texas and Oklahoma. [341]

The reason that the Biden administration moved forward with the listing has little to do with the prairie chicken, itself.

Rather, it has everything to do with the war on fossil fuels. On top of that, it offered the Biden administration an opportunity to attack cattle ranching as well—the latest disfavored industry of the woke left.

The listing is a direct attack on the economy of Kansas, which is home to over 70 percent of the lesser prairie chicken population. It will make it virtually impossible to drill any new oil wells. And it forces ranchers to file annual grazing management plans with a federally designated agency. [342]

This isn't the first time the federal government has attempted to use the prairie chicken as a means to push its climate change agenda. When the Obama administration tried to list the prairie chicken as a threatened species in 2012, oil and gas industry workers were fined up to $45,000 if they disturbed a single acre of lesser prairie chicken habitat. Ranchers could be fined up to $25,000. Fortunately, that listing was defeated in court. [343]

This excerpt shows that little is off the table in the war on fossil fuels.

Politicians and Political Pundits

Panic doesn't just lead us toward bad or ineffective policy solutions – it can also lead us to focus on the wrong problems. [344]

Ultimately, people make decisions based on their own self-interest. For example, a person might buy a vehicle because of its style, gas mileage, environmental footprint, brand loyalty, maneuverability, color, size, or something else.

In contrast, politicians are ostensibly elected to make decisions on behalf of their constituents for the betterment of those constituents. However, it is apparent that many politicians make decisions for the betterment of their own wealth, campaign contributors, and re-election chances. This means that powerful politicians can kill energy-related projects that, if implemented, will compete with their own business interests. [345]

Extreme events have occurred since the Earth was formed approximately 4.5 billion years ago. That event itself was perhaps the most extreme of extreme events. The sudden disappearance of the dinosaurs was another extreme event. Yet many politicians and political pundits do not miss the opportunity to characterize out-of-the-ordinary events such as a river that occasionally overflows, high temperatures for a few days, and a storm that disrupts electric service as proof that we must immediately act to mitigate climate change and global warming.

In addition, politicians can make decisions after hearing only one side of an argument, as can occur when campaign donors and professional lobbyists attempt to influence legislators and other government officials. Political pundits can mislead the public if they have not done their research.

Climate change and global warming are too important to treat as routine. Therefore, people who make decisions on the public's behalf or provide information to the public about climate change and global warming policy should educate themselves on both sides of the issues prior to acting.

Political Action Committees

In the United States, political action committees (PACs) collect contributions from people who ostensibly support their causes for or against candidates for public office, ballot initiatives, and legislation. Some PACs have well-thought-out agendas based on data and definitive plans to apply the funds they raise. However, many PACs raise funds based on much less and distribute them as they please.

For example, consider the statements of a celebrity spokesperson for a PAC whose mission is to "leverage the donations of those who are climate concerned to counter the outsized influence the fossil fuel industry has on our government. We will make sure politicians who support oil and gas are as afraid for their jobs as we are about the impending climate disaster." [346]

The website provides the following dire warning. (**Emphasis added**)

> We have reached a stark turning point. We are no longer just imagining how the world will look in a disrupted climate; we're seeing and feeling the reality of the crisis every single day as we witness wildfires, heat waves, and floods destroy communities. The most recent IPCC report told us that half

of the world population is already in the danger zone, that every fraction of a degree matters, and that every second counts. **Scientists tell us we have just eight years left** to curtail fossil fuel use and prevent the worst climate outcomes. [347]

In a recent interview, its spokesperson said (**emphasis added**):

> Well, you know, you can take anything — **sexism, racism, misogyny, homophobia, whatever, the war** … And if you really get into it, and study it and learn about it and the history of it, everything's connected. **There'd be no climate crisis if it wasn't for racism.** [348]

This effectively says that racism and a litany of other maladies caused the climate crisis. Interestingly, it provides no data to support the link between racism and a climate crisis, which is purportedly only eight years away in 2023.

The Press and Politics

A functioning press thrives on its reporters being fair, accurate, and objective, so their articles can be clear, concise, and coherent. News outlets have tended to lean one way or another, sometimes strongly, since revolutionary times. However, by observation in the early 2020s, news outlets have bent over backwards to support one political party and silence the other. This negatively affects the ability of the press to be fair, accurate, and objective.

The current climate in the press is to amplify the voice of the group with which it agrees. Most of the press supports Malthusian worldview causes, such as climate change, which is high on the list of priorities. Therefore, climate change can, and typically does, have a loud voice in the room. Many supporting the Malthusian worldview promote the notion that climate change is a crisis that needs to be addressed immediately, despite evidence that it needs to be addressed, but not immediately. Further, they imply that only they can fix climate change, so the message is to give them power lest the population die out because of climate change.

The regulations to limit carbon dioxide emissions suggested in Chapter 17 will generally appeal to those holding a Malthusian worldview, who need governmental control to achieve their ends. Stated differently, regulation is the answer that Malthusians want, and that Cornucopians will oppose. If Malthusians have the power to enact legislation, the biggest challenge will be to structure regulations that are not onerous but rather respect the end-user's investments by phasing them in over a long period of time. If Malthusians cannot enact legislation, the biggest challenge may be convincing Cornucopians to address the issue objectively.

Climate change is a scientific issue, not a political issue. Acting in a bipartisan manner based on facts and data will result in the best solution and implementation.

Fake News

Radio, television, and cable media outlets present news stories to the population that are largely relegated to 30 seconds to 3 minutes in length. Print news stories are similarly limited in size. Recognizing that the content of this book cannot be crammed into a 3-minute news report or a short article, it is reasonable to presume that an overwhelming majority of the population does not have a working knowledge of climate change or global warming. Complicating the problem are reporters who do not have an appropriate technical background to understand the various aspects of the subject and their interrelationships.

The result is errors, errors of omission, lies, spin, irrelevant facts, and often tortuous logic to justify the desired conclusion. For example, Figure 10-8 shows only some of the heat wave data, and Figure 11-1 and Figure 11-2 show starkly different temperature data. Another example was presented in Chapter 11, where multiple news outlets erroneously reported that New York City was proposing to regulate carbon emissions from coal-fired and wood-burning pizza ovens when particulates were being targeted.

Maurice Strong, who gave the opening statement to the 1992 Rio Summit, was reported to have dropped out of high school at the age of 14... [349] [350] In contrast, a 10-page article based on an interview said that, when he left his hometown in 1943, having skipped four grades for his scholastic abilities and graduating at age 14... [351] He was born in 1929, so the age of 14 in 1943 is correct. Strong achieved much during his lifetime, so whether he graduated or dropped out is not important now. However, one of these statements fails to correctly characterize Strong as either a high school dropout or a genius. There is a difference.

Keeping this in mind, many widely quoted statements were not used in this book because their original sources could not be located. In general, statements are suspect when they are sensational and utilize tortuous logic. It may not be easy, but **you need to do your own investigation**. Someone saying something, does not make it a fact or the truth. The lack of information about polar bears in Chapter 10 and the discussions in Chapter 11 show that the data and its analysis can be credibly questioned.

Perspective

Putting climate change into perspective requires an extensive background, so it should come as no surprise that virtually everything presented about the subject is incomplete.

For example, high concentrations of carbon dioxide in the atmosphere will suffocate everyone. This can create doubt and fear about the future of humanity. However, reaching 1000 ppm will take 251 years if its rate of increase remains stable. Even if the rate doubles, it will take over a century. However, indoor air quality will be adversely affected sooner. Will humans be able to mitigate this in the next century or so? Probably. Will people who are alive today live for another century? Probably not. Is there concern about the future? Yes. Should people lose sleep tonight over this? No. Should one live in fear? No. [1]

You may have seen photographs of an Arctic polar bear fishing from a small chunk of ice in the Arctic Sea. It was intended to imply that the poor, lonely bear's fishing habitat is diminishing. This same scene could have played out 500 years ago, before humans burned fossil fuels. This is not to say that the bear's fishing habitat is not diminishing. It was. However, the Arctic Sea ice extent increased significantly between 2014 and 2022, when more carbon dioxide was emitted by humans.

The photographs do not mention the date they were taken or exactly where they were taken. Omitting these details lets the viewer fill in the blanks and conclude that the bear is losing its habitat, so

---

[1] Note that this example contains elements of Rosling's gap, negativity, fear, size, destiny, and urgency instincts. Left unchecked, they can play an oversized role in perceiving the importance of an issue and its urgency.

something must be done quickly before it is too late. At least six of the ten Rosling instincts presented in Chapter 2 are at play, including the *negativity, straight-line, fear, size, blame, and urgency instincts*, to elicit an emotional response. At the same time, data shows that the Arctic Sea ice extent is improving.

The photographs of the bear fishing on a piece of ice may or may not be in context, but at least the photograph is real. Nonetheless, it was used to develop a narrative to evoke an emotional response and influence behavior.

People, groups, and political parties have been known to commit errors of omission. For example, electric vehicles are touted as having low emissions, even though it appears that less expensive hybrid vehicles have the same or lower emissions while causing less harm to the environment, not requiring investments in new infrastructure, and not needing to be charged during long trips.

Remember, climate change and global warming are scientific matters, not emotional issues. Start by putting the data into perspective.

Chapter 21: Success Hiding in Plain Sight

Good guys finish last. --- *Attributed to Leo Derocher*

Growing up is a process of expanding circles. A newborn baby's circle encompasses its mother, eating, and its father. As the baby grows, it learns about its ever-expanding circle, which includes its room, house, street, neighborhood, and so on.

Every parent uses a different technique to teach their child, as does every grandparent. Some explain things over and over, hoping that the child will understand, and absorb the material. Others use reverse psychology and sometimes provide wrong answers in such a way that the child can sense that something is not right.

For example, while babysitting, my grandson asked where his parents were. I could have explained that they went out for dinner, but that would have elicited even more questions and anxiety. So, I opened one of my pockets, looked inside, and said that they were not in there. His wheels cranked for a moment before he smiled and returned to playing with his toys.

About eight months ago, my now six-year-old grandson asked why I occasionally give him wrong information. This is an exciting question, so I asked him to *think* about why I do it. Weekly checks revealed that he did not give the question any thought, but we would always discuss it for a while. He usually tried to guess the answer, and I would tell him to *think* about it some more.

After about three months of this, he guessed that I was trying to teach him something. BINGO! He was excited, so I asked what I was trying to teach him. I check back occasionally, only to discover that he has not thought about it. But we still discuss it each time. Either he will get it, or I will tell him, or he will read this story someday.

Truth be told, there are many reasons why I do this, but the primary reason is to teach him to *think for himself* and not accept what other people say, because others may be mistaken, insincere, disingenuous, uninformed, or misinformed.

Another reason is to show him that the answers to many questions and solutions to many problems can be right in front of his nose. You may recall that I repeatedly told him to *think about it* every single time we spoke. In other words, I gave him the answer tens of times, but he did not see it. After all, he is only six years old.

---xxx---

Common knowledge has been disproven many times. For example, for centuries, the Earth was believed to be flat, and the sun, moon, planets, and stars revolved around it. The global cooling of the 1970s [352] morphed shortly thereafter into global warming, before being incorporated into climate change, which is not only more complex and difficult to understand than global warming but interestingly provides more opportunities to trigger Rosling's *fear instinct* and *urgency instinct*.

Has anyone been held to fully account for the 180-degree change in narrative from global cooling to global warming? People say that immediate action is necessary, but is it necessary if a carbon dioxide limit will be reached in approximately 251 years at its present rate of increase or in around a century if the rate of increase doubles?

Of particular concern is the downplaying of Climategate, the apparent suppression of evidence disputing the impact of global warming, and the ostracization and ridicule of people challenging the generally accepted narrative. Why does the IPCC temperature data not include the Medieval Warming Period and Little Ice Age? Were these events global or local? What if the skeptics are correct, and global warming is not as urgent a matter as previously thought? Why does this appear to be playing out under a cloud of doubt?

Policy Development Process

The analysis in this book starts with data, identifies a problem, and suggests technological solutions, while at the same time recognizing that technological solutions may not be sufficient to solve the entire problem. This approach contains elements of both Malthusian and Cornucopian worldviews.

Cornucopians attempt to view the world realistically and develop ways to remove barriers to human progress and prosperity. In contrast, Malthusians view the world as limited, so they promote population reduction and less consumption to avoid catastrophe. However, less consumption would result in a deteriorating standard of living for many.

Both Malthusian and Cornucopian worldviews are hampered by Rosling's *single instinct*. Malthusians limit resource consumption to avoid disaster, whereas Cornucopians analyze problems and apply technology to solve them. Cornucopians have gradually taken charge, starting around 1800. Notwithstanding that Cornucopian accomplishments are ongoing, Malthusians have gained significant influence in the press, schools, international organizations, Western governments, and some businesses.

It is not clear how Malthusians intend to limit resources to sustainable levels and allocate them to countries and communities with different economic and governmental systems. According to The Club of Rome, the world population will be sustainable at a level of well-being roughly equivalent to that of the lower-income European nations. Therefore, some countries would get more than they have, while others will get significantly less, which would effectively constitute a wealth transfer and likely entail some form of governmental intervention.

However, increasing one's standard of living is not just a matter of giving the person money or things, even though the person may very well need them, because one-time transfers of wealth often tend to be spent, effectively leaving the recipient in the same position as before. What they really need is to learn how to generate income and transmit this skill to their children, such as by implementing the five fundamental activities presented in Chapter 4.

Of great importance is education, which is heralded by Cornucopians, who view people as more brains to innovate and more hands to develop economic growth. Urbanization and advances in transportation and communications have provided many educational opportunities that were not available prior to around 1800, including online learning from virtually anywhere in the world via the Internet.

The primary focus of obtaining formal and informal education is usually to provide material support for one's family, often long before that family is formed. Education can also elevate one's self-esteem and enable individuals to achieve in areas that are unrelated to their education, per se. For example, some years ago, a boiler house operator made a request to borrow the company's temperature instruments for a few months. As it turned out, the experts could not mate two nearly extinct birds, so the boiler operator was given the task because of his extensive knowledge in this area. My recollection is that he succeeded where the experts failed.

On a human level, every parent wants their children to succeed. Anecdotally, another boiler operator told me that he was bitter because his father raised him to only do the minimum in school, despite his innate ability to do better. Later, recognizing this father's grave error, he strongly encouraged his daughter's education, so she studied and became an environmental engineer.

Successes and Failures

Occurrences that are Cornucopian in nature generally result in slow change that is boring and usually not reported. Overall, perhaps the largest Cornucopian success was the gradual transition from a Malthusian world to a Cornucopian world starting around 1800, which enabled many in the population to enjoy a much higher standard of living and not bury almost half of their children before they reached the age of 5. This transition was described in the first few chapters of this book and will not be repeated here.

However, Cornucopians also had their failures, including polluting the air, water, and ozone layer, to the extent that Malthusian solutions were imposed to keep pollution within reasonable limits. Similarly, Cornucopians have failed to prevent greenhouse gas emissions from accumulating in the atmosphere.

Malthusians have had considerable success in exposing Cornucopian failures, enacting laws, writing regulations to mitigate pollution, incenting technological improvement, revitalizing the ozone layer, and mitigating air and water pollution.

> Without a doubt, *Limits* became the most powerful scientific paper of the last 50 years. Ever since, uncountable references can be found, numerous actions have been undertaken, technologies have developed, faculties have started, regulation has got off the ground. Resource policies can be traced back to *Limits*, and so can climate and energy innovations. [353]

The accumulation of greenhouse gases in the atmosphere remains to be effectively addressed. Cornucopian solutions have not provided mitigation, so Malthusian environmental regulations have been proposed to address various aspects of this issue.

Malthusians have also had their failures, including the mandate for electric vehicles in California in the 1990s, the inability of mathematical models to agree with actual data, such as illustrated in Figure 11-3, predictions of global cooling in the 1970s followed by global warming in the 1980s, longstanding predictions that the Earth will run out of oil, and the fact that the Earth currently supports approximately 8 times more people than were living when Malthus wrote his essay in 1798, effectively contradicting the Malthusian assertion that food and resources are limited and could not support more people.

Tactics

In general, the tactics employed by both groups have not been all that honorable. Cornucopians often vehemently oppose Malthusian proposals that hamper production and limit development. However, they typically concede when pressured by the population. For example, addressing the accumulation of ozone in the upper atmosphere was opposed until scientists were able to effectively communicate the issue to the population and elicit support that could not be ignored.

Many actions are readily visible, however uncovering action that is not taken often means obtaining an in-depth understanding of the issue, developing approaches to address it, and comparing these approaches to the action taken. This often involves extensive research and intense thought that many people do not have the information necessary to perform, choose not to perform, or cannot perform. Uncovering action not taken takes time and effort, which makes it a powerful tactic.

Groups are often exploited to promote policy. For example, the Committee on Energy and Natural Resources held a hearing on 23 June 1988 to alert policymakers about global warming. Senator Timothy Wirth later told a reporter (**emphasis added**):

> What we've got to do in energy conservation is try to **ride the global warming issue. Even if the theory of global warming is wrong,** to have approached global warming as if it is real means energy conservation, so **we will be doing the right thing anyway** in terms of economic policy and environmental policy. [354]

In 1991, The Club of Rome stated (**emphasis added**):

> **In searching for a new enemy to unite us, we came up with the idea that pollution, the threat of global warming,** water shortages, famine and the like would fit the bill. In their totality and in their interactions these phenomena do **constitute a common threat** which demands the solidarity of all peoples. But in designating them as the enemy, we fall into the trap about which we have already warned, namely mistaking symptoms for causes. All these dangers are caused by human intervention, and it is **only through changed attitudes and behaviour that they can be overcome. The real enemy, then, is humanity itself.** [355]

Both statements display a willingness to exploit global warming for policy reasons. The first statement displays a willingness to do so, even if the theory is wrong. In addition, both statements start with solutions and then seek problems to justify them, instead of the more logical process of defining a problem and then developing solutions. Note the directness and brazenness with which these statements state policy. The first statement was given to a reporter, whereas the second was published. Neither attempted to hide their intent.

Groups sometimes misrepresent facts to sensationalize their cause and achieve their goals, such as in the *Alar-on-apples* scandal. It is worth noting that the total net assets of the organization responsible for the scandal may have been small at the time, but they were almost USD 500 million in 2022, [356] complete with a team of lawyers, scientists, and campaigners [that] are on the frontlines every day waging fierce courtroom battles and hard-hitting campaigns to … advance urgent climate action, [357] which includes climate change and global warming.

> In 1989, the popular television show "60 Minutes," the Natural Resources Defense Council and the actress Meryl Streep denounced this chemical used to regulate the ripening of apples as "the most potent cancer-causing agent in our food supply" and said it was a cause of childhood cancer.

The accusation was based on a 1973 study in which a byproduct of Alar caused tumors in mice. The dosage used in the study was eight times greater than the so-called maximum tolerated dose, the amount above which tissue damage occurs even from innocent substances because of the high concentration.

Subsequent tests by the National Cancer Institute and the Environmental Protection Agency failed to show that Alar caused cancer. Only when mice were given extremely high doses, equivalent to 133,000 to 266,000 times the amount a preschool child might consume in a day in apples and apple juice, did any tumors result.

Still, millions of alarmed parents panicked and dumped untold gallons of apple juice and bushels of apples, the apple industry lost about $375 million, the Department of Agriculture lost another $15 million, countless children were given far less nutritious drinks in place of apple juice and Alar was taken off the market by its manufacturer. [358]

A contemporaneous newspaper article [359] reported (**emphasis added**):

> After this year's stir over use of the chemical Alar on apples, political publicist David Fenton celebrated the work of his firm in a lengthy memo to interested parties. He wrote of a "sea change in public opinion" that has "taken place because of a carefully planned media campaign, conceived and implemented by Fenton Communications with the Natural Resources Defense Council." **Here are a few extracts**:
>
> In the past two months, the American public's knowledge of the dangers of pesticides in food has been greatly increased. Overnight, suppliers of organic produce cannot keep up with demand. Traditional supermarkets are opening pesticide-free produce sections. ...
>
> The campaign was based on NRDC's report ``Intolerable Risk: Pesticides in Our Children's Food." Participation by the actress Meryl Streep was another essential element.

Usually, public interest groups release similar reports by holding a news conference, and the result is a few print stories. Television coverage is rarely sought or achieved. The intensity of exposure created by design for the NRDC pesticide story is uncommon in the non-profit world.

Our goal was to create so many repetitions of NRDC's message that average American consumers (not just the policy elite in Washington) could not avoid hearing it -- from many different media outlets within a short period of time. The idea was for the "story" to achieve a life of its own and continue for weeks and months to affect policy and consumer habits. Of course, this had to be achieved with extremely limited resources.

In most regards, this goal was met. **A modest investment by NRDC re-paid itself many-fold in tremendous media exposure (and <u>substantial, immediate revenue</u> for future pesticide work). In this sense, we submit this campaign as a model for other non-profit organizations.**

This event was destructive and shows that groups can **knowingly misrepresent facts and be willing to use less than ethical means to achieve their goals with little or no concern for the collateral damage** incurred because of their actions. One might consider it reprehensible to suggest the spreading of unsubstantiated claims as a model for other organizations.

Eroding Living Standards

It appears that climate change and global warming may have been, and may still be, manipulated to promote policy. **This begs one to ask the question as to whether similar tactics were used to promote policies in other areas.** Although beyond the scope of this book, many seemingly unrelated events do have an effect. For example, COVID-19 epidemic and the 2020 riots in the United States created acrimony, fear, and conflict in the United States that hindered the normal functioning of society, which reduced the consumption of resources, such as gasoline in Figure 12-1, and lowered the standard of living.

Aside from limiting population growth, the Malthusian approach of limiting global resource consumption, echoing Rosling's *single instinct* of using the same tool for every task, adversely affects the standard of living in those countries with high resource consumption. Conversely, if the standard of living in these countries falls, their population will consume fewer resources. Therefore, the same desired effect can be achieved by limiting consumption, reducing living standards, or both.

In general, activists advocate for their cause by making people aware of it, suggesting legislation, lobbying, protesting, demonstrating, boycotting, and the like. Global environmental successes include the mitigation of air pollution, water pollution, and repairing the ozone layer. Activists have also influenced many other issues, including protecting whales, land conservation, and the construction of factories, pipelines, and other industrial facilities that directly and indirectly reduce population and resource consumption.

At the same time, groups and institutions can implement actions designed to reduce the standard of living in more developed countries to achieve their goals, despite the known collateral damage that these actions cause. **This may seem mean and counterproductive, but it would <u>not</u> be if the goal is to limit natural resource consumption because, *in their opinion*, there is no other way to save mankind.** Stated differently, the cause is existential, so the result justifies whatever means are necessary to achieve that result. It is also an *opinion* that may not be based on facts, so it is possible that a Cornucopian solution might exist or that the proposed Malthusian solution fails to address the underlying cause of the problem and might cause harm instead.

This begs the seemingly impossible question of how to lower the standard of living in societies that consume large amounts of resources and limit population growth in others. The latter can be achieved by helping people move from Level 1 to Level 2, where they will learn that having fewer children results in a better quality of life for their entire family. Billions of people alive today, their parents, their grandparents, or their great-grandparents have made this fundamental transition.

The first step to lowering the standard of living in countries that consume large amounts of resources is to identify the offending countries. Per Figure 19-1, the United States, Canada, China, South Africa, the European Union, and the United Kingdom emit more carbon dioxide per capita than the global average. Other countries may meet this criterion, but the biggest and historically most egregious emissions target is the United States.

If you were given the assignment, how would you go about lowering the standard of living in the United States? Military conquests have been used to subjugate people throughout history, but the United States is a strong country militarily. Actions that impede economic development are another option. A softer approach is to create conflict and acrimony with the intent of progressively weakening society until it breaks down and its people are forced to lower their standard of living and consume less resources. Many such actions may seem innocent on the surface, but in total, they can **create and foment acrimony that divides people, detracts from productive activities that generate wealth, increases non-productive expenditures, and reduces the ability to consume resources. Activities that lower living standards can be destructive on many levels, but they are _successes_ in the eyes of those who implement them in pursuit of their objectives.**

Activism

Activists and community organizers help organize groups and coordinate efforts to promote community interests, such as civil rights, environmental protection, labor, community development, and health care. The process of organizing involves defining and assessing community interests, setting goals, planning actions, developing sustainable leadership, and mobilizing members. It is noteworthy that community organizers are not required to be sympathetic with the cause because their objective is to help guide the organizing process.

Many activist groups are formed by or within larger organizations, such as lobbyists or charities. Grassroots activist groups are formed by ordinary, like-minded people. Activist activities include protests and marches, street campaigning, digital campaigning, political lobbying, putting up posters and distributing flyers, mass action, occupying spaces, and striking. Volunteer activities within the group include training new volunteers, planning events, managing social media, contacting the press, internal communications, and group meetings.

Virtually all activist groups instigate change and therefore create conflict, even when their cause is just. However, the foundation of

266

some groups may be based on false pretexts, be factually disingenuous, play on emotion, cause unnecessary dissent and acrimony, and be designed to lower living standards, such as the groups that advocate discontinuing the burning of fossil fuels prior to the availability of viable alternative energy sources. **What they say is not nearly as important as the effects of their actions or inaction.**

What, When, and How Fast?

In general, it would be prudent to define a problem, determine its root cause, develop solutions that treat the root cause, evaluate the solutions, select the solution that best mitigates the problem (preferably at its root cause), and implement the solution, presuming that it is viable.

Reducing the concentration of carbon dioxide in the atmosphere can be addressed by reducing carbon dioxide emissions, capturing carbon dioxide for permanent storage, or both. Improved processes, renewable energy, conservation, and recycling are woefully inadequate to address the problem at hand, even if more widely implemented. Significantly reducing or eliminating carbon dioxide emissions appears to hinge on increasing near-zero-emission production of electricity and process heat, the technology for which is either not sufficiently available (geothermal, hydropower), not reliable (solar, wind), or unacceptable to society (conventional nuclear). While all have legitimate roles in electricity and process heat production, depending solely on these technologies to replace fossil fuels and meet future needs would not be prudent.

However, there are promising developments to produce near-zero-emission electricity and near-zero-emission process heat, such as small nuclear reactors and carbon capture, utilization, and permanent storage (CCUS). It is not known if or when these (or other technologies) will become viable. However, it is extremely likely that viable technologies will be developed for widespread implementation well before the carbon dioxide concentration in the atmosphere reaches 1000 ppm and starts to affect human health. Therefore, it would be prudent to let the technology develop sufficiently before mandating mitigation.

Implementing mitigation remedies too soon can be expensive, inefficient, wasteful, cause undue hardship, and be potentially inappropriate if significantly improved technology becomes available shortly after installation, whereas late implementation unnecessarily delays mitigation.

Knowing what to do, when to do it, and how fast to do it, is difficult, but that is what is needed. Examples of actions taken at reasonable times include the adoption of LED light bulbs, fuel consumption requirements for vehicles, and furnace efficiency standards. California's mandate requiring automakers to provide electric vehicles, which was repealed in 1996, [360] was enacted decades too soon.

However, taking inappropriate action is a grave concern. For example,

> The 17 Sustainable Development Goals (SDG) are a universal call to action adopted by the United Nations in 2015 to end poverty, protect the planet, and improve the lives and prospects of everyone, everywhere. [361] The Sustainable Development Goals are: no poverty; zero hunger; good health and well-being; quality education; gender equality; clean water and sanitation; **affordable and clean energy**; decent work and economic growth; industry, innovation, and infrastructure; reduced inequalities; sustainable cities and communities; responsible consumption and production; **climate action**; life below water; life on land; peace, justice, and strong institutions; and partnerships for the goals. The SDGs emphasize the interconnected environmental, social, and economic aspects of sustainable development by putting sustainability at their center. [362] **(Emphasis added)**

These are laudable goals that sound great, but findings in this book show that solar and wind energy are not that clean or affordable without subsidies, living standards would be reduced to approximately those of the lower-income European nations to be sustainable, and implementing the Sustainable Development Goals will require a system of social organization in which private property and the distribution of income are subject to social control, which is not compatible with the United States Constitution. It is likely that a critical examination of other Sustainable Development

Goals (SDG) would reveal similar findings. Therefore, implementation may have wide-ranging effects that benefit some groups while having a detrimental effect on others. The challenge is to develop solutions that benefit almost all groups to promote greater good.

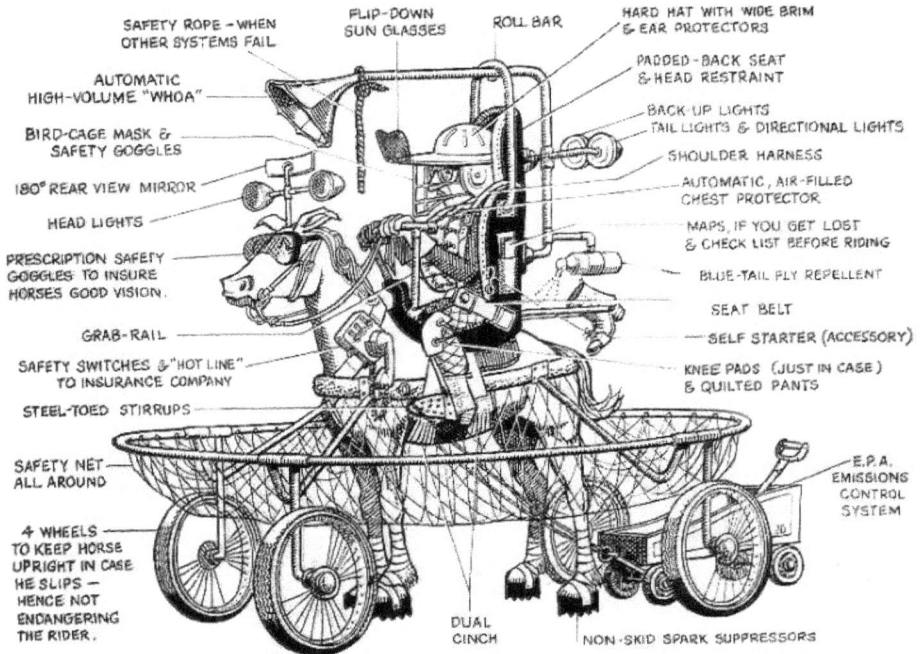

Cowboy after O.S.H.A. Inspection

Figure 21-1 Cowboy After OSHA Inspection

Figure 21-1 is a parody from my files that has been around for decades and takes a humorous look at what can happen when regulations take on a life of their own by satirizing how overzealous regulation can be counterproductive.

Misguided Intelligent People

There are many activist causes that are worthy of support. However, there are many that espouse what are ultimately destructive messages and actions. Both good and bad causes will find a following among the billions of people living on Earth, in part because combinations of Rosling's attributes described in Chapter 2 are often carefully crafted into messages that effectively convince

269

potential members to join and act. Some people will recognize well-disguised destructive messages, whereas others, sometimes referred to as *useful idiots*, will be easily persuaded to believe, say, and do things that are destructive to themselves, those around them, and others.

With so many university-educated people espousing dubious ideas and policies, it appears that the *useful idiots* of the past have been supplanted by *misguided smart people* that practice Rosling's *single instinct* of using the same tool for every task instead of considering multiple tools from which the best can be chosen for the circumstances at hand.

For example, approximately half of the United States Congress in the mid-2020s espouses higher taxes, bigger government, and defunding the police to solve problems, while the other half advocates for lower taxes, smaller government, and police enforcement. Depending on which half you ask, high inflation, increasing debt, and increasing crime are either the result of taxes not being high enough, government not being big enough, and too many police, or taxes being too high, government being too big, and not enough police. Interestingly, these diametrically opposing conclusions were reached by university-educated people who have access to essentially the same information.

Perspective

There are a multitude of counterproductive, divisive, and destructive actions and events that tend to limit the population. its consumption of resources, and its standard of living. Conversely, there are thousands upon thousands of productive occurrences that are not reported because they are small, hard to understand, boring, and individually inconsequential but collectively important, that improve the lives of many. Stated differently, we are surrounded by Malthusian and Cornucopian successes and failures.

However, **our worldview tends to influence if we see them, how we see them, how we interpret them, and what we do about them.** The proverb 'beauty is in the eye of the beholder' personifies this thought. Despite seemingly noble intentions, advocates for many diverse causes often support actions and events that adversely affect people. Therefore, **support for a particular cause should be**

evaluated <u>not</u> by its intentions, but rather by how its actual or proposed actions affect the population.

Chapter 22: The Consequences of Solutions

Don't bite off your nose to spite your face. --- Popular Metaphor

Does a fish know that it is wet? Maybe not, if the fish was in water all its life and never had the experience of being out in the air. Or maybe the fish knows. Feel free to ponder that thought.

Some years ago, an assignment took me overseas for a year. I was excited because I had never been overseas. Where? It did not matter because it was *overseas*. My assignment was in a country that had a sizable Level 1 population but where most people lived at Level 2 and Level 3. Immigrants from other countries were present, but there were not nearly as many as in my native country.

My boss gave me three pieces of advice before sending me off: take lessons to learn the local language, get a car, and do not sit in the hotel bar drinking beers. Upon arrival, I went to a local English school, arranged for local language classes, bought a car, and do not recall ever sitting around drinking beer in the hotel bar.

Having caught a bit more than a cold during my second weekend onsite, not knowing any medical professionals, and coming from a Level 4 country, I walked down to the local hospital to visit a doctor. I also saw people in much worse shape who were waiting to be seen. I never went back during the year that I was assigned there, so others could have my turn.

Everyone treated me especially well because treating guests well is part of their culture. My effort to learn their language and not being bashful about speaking with an accent did not hurt. I had no choice but to accept the treatment, but I did find it uncomfortable.

Nonetheless, I learned more about myself and my native country by living there than they could ever imagine, because I was in a different fishbowl.

---xxx---

The world was Malthusian prior to around 1800, before technology and industry started advancing and the world became increasingly Cornucopian in the quest to increasingly address human needs. Human activity intensified and ran into limits in the mid-1900s that were highlighted by Malthusians. Pollution and other issues were subsequently addressed in the late 1900s. By the 2020s, Malthusians had achieved dominance in the press, schools, environmental organizations, Western governments, international organizations, and some businesses.

This might suggest that Cornucopians have been asleep at the wheel. However, technological improvements and more efficient utilization of resources, which are Cornucopian in nature, are collectively important, occur daily, but are not reported because they are relatively small, boring, and happen all the time. Cornucopian actions that fail are typically small, easily corrected or abandoned, and usually collectively unimportant.

In contrast, Malthusians tend to have fewer achievements, which are generally of much larger magnitude, such as research, conferences, legislation, and regulations that can have significant positive or negative effects on the population. **Avoiding the negative effects is of extreme importance** because they can be expensive, devastating to the population, or both. For example, spending an exorbitant amount of time and money on solar geoengineering projects attempting to reduce the temperature of the Earth will not address the problem of increasing carbon dioxide concentration in the atmosphere, aside from potentially damaging the Earth in unintentional ways.

Most individuals have no interest in destructive narratives and do not knowingly embrace them. However, a vocal minority, knowingly or unknowingly, is interested and does embrace them. Partial immunization from being swept away by destructive narratives can be achieved by understanding where the narratives lead. It is suggested that one take the time to agree or disagree with the following series of statements, which might provide some clarity.

Malthusian and Cornucopian Worldview Statements

Perhaps the easiest distinction that can be made about oneself is the tendency to support the Malthusian worldview, Cornucopian worldview, or both. The Malthusian worldview is correct in the sense that there are limits to what humans can do to the Earth without having adverse effects. Consuming the Earth's resources responsibly makes sense. The Cornucopian worldview has been shown to be correct in the sense that limits have been overcome using technology. For example, the Earth will sustain approximately 8 billion people in the 2020s, as compared to approximately 1 billion much less prosperous people in 1798, when Malthus published his essay.

Both worldviews appear to be technically correct. However, Rosling's *single attribute* of using the same approach to every problem hampers their respective solutions. Here, Malthusians suggest mitigating global warming by reducing population and limiting resource consumption, whereas Cornucopians suggest applying technology to solve problems. Both solutions are flawed because limiting resource consumption will drastically reduce standards of living, and technology may not be capable of mitigating the rising concentration of carbon dioxide in the atmosphere and global warming.

The following statements may help discover whether one tends to be Malthusian, Cornucopian, or something else. These statements and those that follow are intended to evoke thought but are not intended to be essay topics, so one should either agree or disagree with each of them.

- Earth's resources are scarce, so resource consumption needs to be limited for humans to survive.
- People must always be kept down to the level of the means of subsistence. [363]
- Solutions that restrict resource consumption should be implemented first, and if they do not work, technological solutions should be tried.
- Technological progress will meet the needs of a growing population, so it needs to be fervently pursued.
- People represent more brains to innovate and more hands to develop economic growth.
- Technological solutions should be implemented first, and if they do not work, resource consumption solutions should be tried.

You might agree with all of them to varying degrees, but the ones with strong agreement might help identify one's leanings. Malthusians would agree with the first three statements, while Cornucopians would agree with the last three. A Malthusian pragmatist would agree with the third statement, and a Cornucopian pragmatist would agree with the sixth.

That said, these statements question underlying principles and beliefs but do not examine the consequences of the solutions stemming from them. In other words, they may provide some insight, but do not appear to be the correct statements to analyze.

Emotion Statements

Here are some additional statements to consider that are not intended to be essay topics, so one should either agree or disagree with each of them.

- I feel sorry for people who have less than me.
- I live in an affluent country, so everyone should be entitled to enter my country and live here permanently.
- I am willing to significantly sacrifice my standard of living and that of my family for the greater good of the global population.
- I am envious of people who have more than me.
- I live in a less affluent country, so I should be able to move to a more affluent country and live there permanently.
- I want to improve my standard of living and that of my family.

These statements are intended to be personal. The first three shed light on compassion for others and altruism, and the last three concern aspirations, all of which factor into decision-making processes. These statements query underlying beliefs. They may provide insight, but they do not examine the consequences of the solutions stemming from them and do not appear to be the correct statements to analyze.

Global Warming Statements

These statements might help gauge one's thoughts about global warming.

- The increase in the concentration of carbon dioxide in the atmosphere is real.
- Global warming is real.
- Global warming requires immediate action.
- The global population should be reduced, or at least, stabilized.
- The consumption of resources should be limited.
- Technology can mitigate the effects of global warming.

Before writing this book, I would have agreed with all six statements. My answers are more nuanced after investigating the subject. You may agree or disagree with some or all my answers, but the increase in the concentration of carbon dioxide is real, global warming is probably real and warrants action but not immediate action, the global population will stabilize, resources should be consumed wisely, and technology can provide many solutions, so it should be considered before limiting the consumption of resources. Again, these statements might provide insight, but they do not examine the consequences of the solutions stemming from them and do not appear to be the correct statements to analyze.

Cornucopian Worldview Statements

Here are some statements that shed light on the consequences of Cornucopian solutions, which focus on utilizing technology to overcome constraints. These statements largely follow their order of appearance in Chapter 3. These are not intended to be essay topics, so one should either agree or disagree with each of them.

Water, Food, and Sewage
- My family wants to fetch unpurified water in buckets every day from a stream for cooking and cleaning instead of getting water from a sink.
- My family wants to store food in a root cellar instead of a refrigerator.
- My family wants to gather firewood for cooking and heat.
- My family wants to cook on a stove that burns wood or dung instead of a natural gas stove, electric stove, or microwave oven.
- My family wants to grow its own food, raise its own animals, and barter with neighbors for other items instead of buying them at a supermarket.
- My family wants to eat all meals at home and none in restaurants.
- My family wants to use an outhouse or latrine instead of a bathroom with indoor plumbing.

Personal Hygiene and Health Care
- My family does **not** do routine medical and dental examinations.
- My family members do **not** wash their hands after urinating and defecating.
- My family members do **not** wash their hands before preparing food.
- My family members do **not** wash their hands before eating food.
- My family wants to clean themselves with a wet sponge instead of taking a bath or shower.

Electricity, Communications, Transportation, and Education
- My family wants to live **without** electricity.
- My family only wants to communicate personally with others instead of using a telephone, smart phone, or computer.
- My family wants to be entertained by people and mechanical toys instead of radio, television, films, videogames, social media, streaming media, and electronic toys.
- My family wants to travel on foot or by horse and buggy instead of by car, bus, train, or airplane.
- My children's education will be at schools within walking distance of my house, despite other schools offering better opportunities.

Disagreement with these statements indicates that one is enjoying the fruits of Cornucopian successes that have been constructive in improving the human condition. It is likely that most people fall into this category. This does not mean that one supports a Cornucopian worldview.

Malthusian Worldview Statements

Here are some statements that shed light on the consequences of Malthusian solutions, which focus on limiting growth, reducing consumption, and limiting the population to overcome constraints. These are not intended to be essay topics, so one should either agree or disagree with each one.

- If the population grows excessively, people will starve for lack of resources resulting in widespread poverty and degradation.
- The population must be kept down to the level of subsistence to survive.

Agreement with the first statement acknowledges the reality that people will starve if they have insufficient food and shelter. There are a host of reasons to disagree with the second statement, if for none other than it being socially unacceptable to advocate keeping people down. Nonetheless, agreement with the second statement is both logical and benevolent when made in a vacuum of not considering an alternative Cornucopian worldview. However, Malthusian leanings might be indicated when there is agreement with the second statement after fully considering the Cornucopian worldview and its successes.

The Bottom Line

Malthusians and Cornucopians are more readily distinguished by the actions they support, not by how they answer questions, the causes they champion, or their political leanings. For example, one of the United Nations' Sustainable Development Goals (SDG) is *no poverty*. [364] Virtually everyone wants to eradicate poverty, but this objective cannot be realistically achieved, if not due to individual abilities and life's challenges, but also because the poverty threshold is periodically adjusted, effectively transforming the classification of poverty into a moving target.

Spending a reasonable amount of money to provide effective resources that help people in need is laudable. However, there are many people and groups who, given the opportunity, would spend eye-popping amounts of other people's money (OPM) on programs to eradicate poverty. Exorbitant spending would effectively divert significant funds from the population, lower their standard of living, and reduce their consumption of resources. This solution effectively advances the Malthusian objective of reducing the standard of living to reduce resource consumption but not eradicate poverty.

Solutions that advance Malthusian objectives often benefit one group to the detriment of another. For example, regulating the efficiency of residential natural gas furnaces that sell at a premium of approximately 10 percent might reduce natural gas consumption and carbon dioxide emissions by approximately 30 percent. However, a high efficiency furnace would not be needed in locations where it will not run often, such as in a vacation home that is occupied only during the summer. Nonetheless, this regulation is reasonable because most people will use the furnace during cold weather, such that the energy savings of the high efficiency furnace will offset the additional cost premium in a relatively short time and provide substantial energy savings thereafter, while emitting significantly less carbon dioxide. This regulation advances Malthusian objectives of consuming less resources, but it is reasonable because it benefits an overwhelming majority of furnace installations while imposing a relatively small cost premium on people who do not benefit as much.

Both Malthusian and Cornucopian solutions to a problem can be good, bad, or somewhere in between. The challenge is to know which solutions are overwhelmingly good and worthy of avid support, and which solutions are bad and should be staunchly opposed. Note that the nature of the problem itself is irrelevant. It is the effect of the solution that is important.

Sometimes, actions that benefit a small group to the detriment of a larger group create division, animosity, and fear that can impair the normal functioning of society and lower its standard of living. For example, the *Alar-on-apples* scandal frightened people into throwing away the apples in their refrigerators and removing a healthy food option from their and their children's diet, lowering their standard of living, which supported a Malthusian objective. The United States Department of Agriculture said Washington apple growers lost at least $125 million in the six months after the initial uproar [365] while the scandal generated "substantial, immediate revenue" [366] for the National Resources Defense Council (NRDC). In short, the scandal benefited few, penalized many, and supported a Malthusian objective.

Determining whether an action, event, or proposed solution supports a Malthusian objective often requires significant effort. The following questions and thoughts might help guide this process.

- What are the positives? They should be easy to find, well-described, and sound great.

279

- Do negatives exist? They almost always exist, such as Climategate.
- Find and investigate the negatives. Locating applicable facts, data, papers, studies, reports, books, and articles will usually require significant effort. To save time and effort, seek out others who may have already done this work.
- Ask for the data that supports positive and negative findings. Question people and groups that do not provide complete and organized data in a timely manner to support their statements.
- Check the data, methodology, and conclusions for accuracy and reasonableness.
- Think carefully about accepting conclusions with little or no supporting data. You may have no choice but to locate the data and analyze it yourself, which can be a lot of work.
- Identify and evaluate the negatives specifically for their propensity to lower the standard of living. Recognize that a person or group may not recognize that their noble cause can result in a counterproductive solution.
- Evaluate whether the positives outweigh the negatives, and to what extent they do so.

By observation, both Malthusian and Cornucopian solutions have been successful. Since 1800, Cornucopian solutions have successfully raised the standard of living of billions of people, while Malthusian solutions have successfully restricted consumption and addressed air pollution, water pollution, and the ozone layer. Given their goal of limiting consumption, Malthusian solutions may be destructive by design, but they are often necessary when no Cornucopian solution exists.

The bottom line is that a person's or group's goals may be extremely laudable, but the real-world consequences of their solutions should be carefully evaluated. Remember the metaphor, *don't bite off your nose to spite your face*.

Epilogue

The further forward we go, the further back we have to explore in order to go forward again. --- Stephen Gardiner

Over two centuries have passed since Malthus wrote about the population not being able to grow beyond that for which its resources can support it, stating that "an obvious truth (is that the) ... population must always be kept down to the level of the means of subsistence." [367] Sustainable development is derived from a Malthusian desire to limit the use of fossil fuels, promote solar and wind energy, and advocate for population reduction.

In contrast, the Cornucopian worldview is that technological progress is the key to meeting the needs of a growing population. People are viewed as more brains to innovate and more hands to develop economic growth. Higher prices for scarce resources incentivize conservation, while new products and processes use scarce resources more efficiently or eliminate the need for them altogether.

The debate over climate change and global warming is ultimately another front in the battle between the Malthusian and Cornucopian worldviews. Malthusian limitations have materialized, while Cornucopian predictions of improving the human condition have largely prevailed.

It does not matter whether one tends to support a Malthusian worldview, Cornucopian worldview, or something else. Words are cheap, but solutions speak volumes. **The ability to think critically and clearly understand the consequences associated with proposed solutions is of paramount importance to improving the human condition, independent of whether the solution is Malthusian or Cornucopian.**

Table of Contents and Subsections

Endnotes

Note: Due to the dynamic nature of the Internet, the location of some items cited in this work that were accessible at the time of writing may change or may not be available.

[1] How Old is the Earth?, American Museum of Natural History accessed on 23 March 2023

[2] History of Air Pollution, United States Department of Environmental Protection (EPA) accessed on 5 March 2023

[3] Factfulness: Ten Reasons We're Wrong About the World and Why Things Are Better Than You Think, Hans Rosling, Flatiron Books, 2018, Pages 47-48

[4] A Brief History of Ozone, Kasha Patel, NASA, released 30 September 2014 and accessed on 5 March 2023

[5] Ozone layer recovery is on track, due to success of Montreal Protocol, United Nations News, 9 January 2023 accessed on 5 March 2023

[6] The Rapid Transition to Energy Efficient Lighting: An Integrated Policy Approach, The United Nations Environment Programme, 2013

[7] State Air Board Repeals Mandate for Electric Cars, Marla Cone, Los Angeles Times, 30 March 1996 accessed on 6 April 2023

[8] Data, Meriam-Webster Dictionary accessed on 15 August 2023

[9] Trump in court for $250 million NY fraud trial that threatens his real estate empire, Kevin Breuninger, CNBC, 7 December 2023 accessed on 30 December 2023

[10] New York AG asks judge to fine Trump $370M and bar him for life from NY real estate industry, Peter Charalambous, Aaron Katersky, ABC News, 5 January 2024 accessed on 6 January 2024

[11] Donald Trump found liable for fraud in New York civil case, Jonathan Stempel and Karen Freifeld, Reuters, 27 September 2023 accessed on 8 November 2023

[12] A judge found Trump committed fraud in building his real-estate empire. Here's what happens next, Linsay Whitehurst, Bernard Condon, Associated Press, updated 27 September 2023 accessed on 12 January 2024

[13] Donald Trump found liable for fraud in New York civil case, Jonathan Stempel and Karen Freifeld, Reuters, 27 September 2023 accessed on 8 November 2023

[14] Trump fraud trial: NY AG seeks $370M fine, NY real estate ban against Trump, Peter Charalambous, Aaron Katersky, Olivia Rubin, Lucien Bruggeman, ABC News, 5 January 2024 accessed on 6 January 2024
Last Updated: January 5, 2024,

[15] 1620 S Ocean Blvd, Palm Beach, FL 33480, Zillow accessed on 28 December 2023

[16] Mar-a-Lago, Wikipedia accessed on 28 December 2023

[17] The Ironic History of Mar-a-Lago, Smithsonian Magazine, November 2017 accessed on 30 December 2023

[18] How Much Has Trump Made From Mar-A-Lago, His Palm Beach Estate Under Siege?, Dan Alexander, Forbes, 9 August 2022 accessed on 2 January 2024

[19] Factfulness: Ten Reasons We're Wrong About the World and Why Things Are Better Than You Think, Hans Rosling, Flatiron Books, 2018

[20] Climate change is killing people, but there's still time to reverse the damage, Rebecca Hersher, updated 28 February 2022, National Public Radio Morning Edition (WAMO) accessed on 17 May 2023

[21] 'The world is going to end in 12 years if we don't address climate change,' Ocasio-Cortez says, USA Today, 22 January 2019

[22] Fourth National Climate Assessment, Volume II, Impacts, Risks and Adaptations in the United States, US Global Change Research Program, 2018

[23] Power Policy --- Plan or Panic?, Bulletin of the Atomic Scientists, May 1972. On a side note, the link to this document located this document, however links to many articles with similar predictions did not work.

[24] India's Transition to Renewable Energy for Sustainable Development, Sharada Prahladrao, ARC Advisory Group e-mail received on 12 June 2023

[25] Urbanization over the past 500 years, 1500 to 2016, Our World in Data downloaded on 22 February 2023

[26] The Great Horse Manure Crisis of 1894, Ben Johnson, Historic UK downloaded on 5 May 2023

[27] How Early Americans Took Care of Their Teeth How Early Americans Took Care of Their Teeth, Dental Express, 7 November 2017

[28] Keep it clean: The surprising 130-year history of handwashing, Amy Fleming, The Guardian accessed on 18 March 2020

[29] About Handwashing, History, Global Handwashing Partnership, History accessed on 21 February 2023

[30] About Handwashing, History, Global Handwashing Partnership, History accessed on 21 February 2023

[31] The Electric Light System - Thomas Edison National Historical Park, U.S. National Park Service accessed on 21 February 2023

[32] Celebrating the 80th Anniversary of the Rural Electrification Administration, USDA accessed on 21 February 2023

[33] *History and Evolution of Public Education in the US*, Center on Education Policy, George Washington University, 2020

[34] Excepted from Table 104.10. Rates of high school completion and bachelor's degree attainment among persons age 25 and over, by race/ethnicity and sex: Selected years, 1910 through 2019, National Center for Educational Statistics accessed on 20 March 2023

[35] Average years of schooling, Our World in Data downloaded on 21 March 2023

[36] Racism and anti-Racism in the World: before and after 1945, Kathleen Brush, 2020

[37] Can Tourist Get Access to 'Japanese Only' Restaurants?, JapanTruly, 4 November 2022 accessed on 29 June 2023

[38] Reparations for All or None, Kathleen Brush, 2022

[39] Liberian nationality law, Wikipedia accessed on 30 June 2023

[40] 1854: No Irish Need Apply, Mark Bulik, New York Times, 8 September 2015 downloaded on 29 June 2023

[41] Reparations for All or None, Kathleen Brush, 2022

[42] Racism and anti-Racism in the World: before and after 1945, Kathleen Brush, 2020

[43] Exodus 21, King James Version, Bible Gateway accessed on 6 October 2023

[44] Death and Dying, Drew Gilpin Faust, National Park Service accessed on 26 June 2023

[45] Reparations for All or None, Kathleen Brush, 2022

[46] 10 Facts About The Arab Enslavement Of Black People Not Taught In Schools, A. Moore, Atlanta Black Star, 2 June 2014, updated on 9 February 2019 and accessed on 26 June 2023

[47] Reparations for All or None, Kathleen Brush, 2022

[48] *An Essay on the Principle of Population*, Thomas Malthus, 1798, Page vii

[49] "Malthusian", Meriam-Webster.com Dictionary accessed on 22 February 2023

[50] "Thomas Malthus", Britannica accessed on 22 February 2023

[51] *An Essay on the Principle of Population*, Thomas Malthus, 1798, Page 4-5

[52] *An Essay on the Principle of Population*, Thomas Malthus, 1798, Page 11

[53] List of wars by death toll, Wikipedia accessed on 22 February 2023

[54] List of epidemics, Wikipedia accessed on 22 February 2023

[55] The size of the world population over the last 12,000 years, Our World in Data downloaded on 22 February 2023

[56] Global Child Mortality, Our World in Data downloaded on 22 February 2023

[57] Mortality rates of children over the last two millennia, Our World in Data downloaded on 22 February 2023

[58] Life expectancy, 1770 to 2021 (graph), Our World in Data accessed on 7 March 2023

[59] Average Number of Babies per Woman from 1800 to Today, Gapminder downloaded on 22 February 2023

[60] Maternity mortality ratio, 1751 to 2020, Our World in Data downloaded on 22 February 2023

[61] The History of Unions in the United States: Milestones in the struggle to protect workers' rights, Ronni Sandroff, Investopedia downloaded on 23 January 2023

[62] Bureau of Labor Statistics News Release USDL-23-0071, 19 January 2023

[63] Child mortality rate (under five years old) in the United States, from 1800 to 2020, Gapminder downloaded on 24 February 2023

[64] Fertility rate: children per woman, Our World in Data downloaded on 24 February 2023

[65] *China's population is shrinking. The impact will be felt around the world*, Jessie Jeung, CNN, 19 January 2023

[66] China's population declines for second straight year as economy stumbles, Laura He, Simone McCarthy, CNN, updated 17 January 2024 accessed on 22 January 2024

[67] *The world in 2100*, United Nations accessed on 24 February 2023

[68] Fertility rate: children per woman including UN projections, 1950 to 2100, Our World in Data downloaded on 13 March 2023

[69] Visualizing the World's Population by Age Group, Carmen Ang, Visual Capitalist, 16 June 2021 accessed on 26 February 2023

[70] Population in the world (2020), Worldometer accessed on 26 February 2023

[71] World Population: Past, Present and Future, Worldometer downloaded on 26 February 2023

[72] World Population; Past, Present and Future, Worldometer downloaded on 14 April 2023

[73] Reparations for All or None, Katherine Brush, 2022, Page 326

[74] Factfulness: Ten Reasons We're Wrong About the World – and Why Things Are Better Than You Think, Hans Rosling, Flatiron Books, 2018

[75] ibid, Factfulness: Ten Reasons We're Wrong About the World – and Why Things Are Better Than You Think, Hans Rosling, Flatiron Books, 2018, Page 33

[76] Share of population living in extreme poverty, World, 1820 to 2018, Our World in Data downloaded on 27 February 2023

[77] Global primary energy consumption by source, Our World in Data downloaded on 2 March 2023

[78] Fossil fuel, dictionary.com accessed on 28 April 2023

[79] Number of coal mining fatalities in the United States from 1900 to 2022, Statistica accessed on 31 October 2023

[80] They went hunting for fossil fuels. What they found could help save the world, Laura Paddison, CNN, 29 October 2023 accessed on 31 October 2023

[81] Methane Reforming: Solving the Hydrogen Blues, Gerald Ondrey, Chemical Engineering, October 2023, Page 13

[82] How much carbon dioxide is produced when different fuels are burned?, American Geosciences Institute downloaded on 29 March 2023

[83] Carbon footprint, Wikipedia, accessed on March 4, 2020

[84] Table 8.1. Average Operating Heat Rate for Selected Energy Sources, 2011 through 2021 (table), United States Energy Information Administration accessed on 7 March 2023

[85] How much electricity is lost in electricity transmission and distribution in the United States?, Energy Information Administration accessed on 7 March 2023

[86] op. cit., Table 8.1. Average Operating Heat Rate for Selected Energy Sources, 2011 through 2021 (table), United States Energy Information Administration accessed on 7 March 2023

[87] U.S. energy consumption by source and sector, 2021, U. S. Energy Information Administration downloaded on 9 March 2023.

[88] https://www.msichicago.org/explore/whats-here/exhibits/coal-mine/ downloaded on 10 March 2023

[89] A Graphical History of Atmospheric CO_2 Levels Over Time, Owen Mulhernaug, 12 August 2020 downloaded from earth.org on 31 March 2023

[90] Ancient Deepsea Shells Reveal 66 Million Years Of Carbon Dioxide Levels, Leslie Lee, Texas A&M Today, 14 June 2021 downloaded on 21 March 2023

[91] Carbon Dioxide Levels in the Atmosphere (80,000 years), Our World in Data downloaded on 13 March 2023

[92] 10,000 Years of Carbon Dioxide, Berkeley Earth downloaded on 13 March 2023

[93] Global atmospheric CO_2 concentration, National Oceanic and Atmospheric Administration (NOAA) downloaded on 17 March 2023

[94] Trends in atmospheric concentrations of CO_2 (ppm), CH4 (ppb) and N2O (ppb), between 1800 and 2017, European Environmental Agency accessed on 26 March 2023

[95] Trends in atmospheric concentrations of CO2 (ppm), CH4 (ppb) and N2O (ppb), between 1800 and 2017, European Energy Agency, 5 December 2019 accessed on 31 May 2023

[96] Trends in Atmospheric Carbon Dioxide (1959-2022), Mauna Loa, Hawaii, Earth System Research Laboratories, Global Monitoring Laboratory, NOAA accessed on 31 May 2023

[97] Mauna Loa Monthly Averages, National Oceanic and Atmospheric Administration (NOAA) Global Monitoring Laboratory downloaded on 22 March 2023

[98] Mauna Loa Monthly Averages 2021 to 2023, NOAA Global Monitoring Laboratory downloaded on 22 March 2023

[99] Annual CO₂ Emissions, Our World in Data downloaded on 22 March 2023

[100] How do human CO_2 emissions compare to natural CO_2 emissions?, CO_2 Human Emissions, 22 November 2017 accessed on 10 July 2023

[101] Trends in atmospheric concentrations of CO_2 (ppm), CH4 (ppb) and N2O (ppb), between 1800 and 2017, European Energy Agency, 5 December 2019 accessed on 31 May 2023

[102] Trends in Atmospheric Carbon Dioxide (1959-2022), Mauna Loa, Hawaii, Earth System Research Laboratories, Global Monitoring Laboratory, NOAA accessed on 31 May 2023

[103] Global CO_2 emissions from fossil fuels (1800-1955), Hannah Ritchie and Max Roser, Our World in Data downloaded on 31 May 2023

[104] Effects of Rising Atmospheric Concentrations of Carbon Dioxide on Plants, Daniel R. Taub, Nature Education Knowledge, 2010 accessed on 18 April 2023

[105] Trends in atmospheric concentrations of CO_2 (ppm), CH4 (ppb) and N2O (ppb), between 1800 and 2017, European Energy Agency, 5 December 2019 accessed on 31 May 2023

[106] Trends in Atmospheric Carbon Dioxide (1959-2022), Mauna Loa, Hawaii, Earth System Research Laboratories, Global Monitoring Laboratory, NOAA accessed on 31 May 2023

[107] Global CO_2 emissions from fossil fuels (1800-1955), Hannah Ritchie and Max Roser, Our World in Data downloaded on 31 May 2023

[108] Greening of the Earth Mitigates Surface Warming, NASA, 23 November 2020 accessed on 19 May 2023

[109] Carl Edward Rasmussen, Atmospheric Carbon Dioxide Growth Rate, University of Cambridge, 6 March 2023 accessed on 22 March 2023

[110] Greenhouse Gas, Wikipedia accessed on 24 March 2023

[111] Annual Greenhouse Gas Index, globalchange.gov downloaded on 24 March 2023

[112] Atmospheric methane, Wikipedia accessed on 24 March 2023

[113] Methane and climate change, International Energy Agency accessed on 3 November 2023

[114] Greenhouse Gas, Wikipedia accessed on 24 March 2023

[115] Airborne Nitrogen Dioxide Plummets Over China, Earth Observatory, NASA downloaded on 18 March 2023

[116] https://commons.wikimedia.org/wiki/File:Atmospheric_Transmission-en.svg downloaded on 24 March 2023

[117] Altermatt Lecture: The Solar Spectrum, 4.3: Atmospheric absorption - an overview, PV Lighthouse downloaded on 23 March 2023

[118] Wikimedia Commons, all_palaeotemps.png downloaded on 23 March 2023

[119] Global temperature change over the last 2019 years, Ed Hawkins, Climate Lab Book downloaded on 23 March 2023

[120] Greenhouse Gases, Richard P Tuckett, ResearchGate downloaded on 23 March 2023

[121] Global cooling, Wikipedia accessed on 7 July 2023

[122] Climate Change: Global Temperature, NOAA climate.gov downloaded on 26 March 2023

[123] Greenhouse Gases, Richard P Tuckett, ResearchGate downloaded on 23 March 2023

[124] Heat Wave Characteristics in the United States by Decade (1961-2021), US Environmental Protection Agency downloaded on 30 March 2023

[125] Heat Wave Characteristics in the United States by Decade (1961-2021), US Environmental Protection Agency downloaded on 30 March 2023

[126] U.S. Annual Heat Wave Index, 1895–2021, US Environmental Protection Agency downloaded on 30 March 2023

[127] Climate Change Reconsidered II: Fossil Fuels (2019), Nongovernmental International Panel on Climate Change (NIPCC), Chapter 7.4.1 accessed on 27 May 2023

[128] Post-Glacial Sea Level.png, Wikimedia Commons downloaded on 26 March 2023

[129] Millennia of sea-level change, Stefan, 22 February 2016, RealClimate downloaded on 26 March 2023

[130] Climate Change Indicators: Sea Level, US EPA downloaded on 26 March 2023

[131] New Data Reveal Stunning Acceleration of Sea Level Rise, John Upton, Scientific American, 22 February 2016 downloaded on 30 May 2023

[132] What is Ocean Acidification?, NOAA accessed on 27 March 2023

[133] Ocean acidity over the past 25 million years and projected to 2100, European Environment Agency downloaded on 27 March 2023

[134] Ocean Acidification Graph, Smithsonian downloaded on 27 March 2023

[135] U.S. and Global Mean Temperature and Precipitation, Exhibit 6, US Environmental Protection Agency downloaded on 27 March 2023

[136] Interactive Sea Ice Graph (Arctic), National Snow & Ice Data Center downloaded on 28 March 2023

[137] 1850 Arctic Sea Ice Extent, Florence Fetterer, Carbon Brief posted on 11 August 2016and downloaded on 28 March 2023

[138] Interactive Sea Ice Graph (Antarctic), National Snow & Ice Data Center downloaded on 28 March 2023

[139] Graphing polar bear population estimates over time, Polar Bear Science posted on 18 February 2014 downloaded on 28 March 2023

[140] Latest global polar bear abundance 'best guess' estimate is 39,000 (26,000-58,000), Polar Bear Science posted on 26 March 2019 downloaded on 28 March 2023

[141] Climate Change Indicators: Sea Surface Temperature, Figure 1, US Environmental Protection Agency downloaded on 11 June 2023

[142] Cumulative CO_2 emissions by world region, Our World in Data downloaded on 12 June 2023

[143] Global fossil fuel consumption, Our World in Data downloaded on 12 June 2023

[144] Ocean Conveyor Belt, National Geographic downloaded on 27 March 2023

[145] U.S. and Global Mean Temperature and Precipitation, Exhibit 6, US Environmental Protection Agency downloaded on 27 March 2023

[146] Tornedos by Decade 1950-2018, National Weather Service downloaded on 28 March 2023

[147] Was 2020 a Record-Breaking Hurricane Season? Yes, But. . ., Chris Landsea and Eric Blake, National Weather Service posted on 30 June 2021 and downloaded on 27 March 2023

[148] Climate Change Indicators: Wildfires, US Environmental Protection Agency updated 22 July 2022 and downloaded on 28 March 2023

[149] Decades of mismanagement led to choked forests — now it's time to clear them out, fire experts say, Alicia Victoria Lozano, NBC News, 18 October 2020 accessed on 9 June 2023

[150] Climate Change Indicators: Wildfires, US Environmental Protection Agency updated 22 July 2022 and downloaded on 28 March 2023

[151] Global Total Wildfire Carbon Emissions, Copernicus Atmosphere Monitoring Service downloaded on 28 March 2023

[152] Safety Data Sheet, Carbon Dioxide, Airgas, Air Liquide accessed on 16 May 2023.

[153] Carbon Dioxide Health Hazard Information Sheet, FSIS Environmental Safety and Health Group (ESHG), US Department of Agriculture accessed on 7 June 2023

[154] Mama Mia! NYC rules crack down on coal-, wood-fired pizzerias — must cut carbon emissions up to 75%, Carl Campanile and Kevin Sheehan, New York Post, 25 June 2023 accessed on 26 June 2023

[155] The New York City pizza stove controversy, explained, Ben Adler, 30 June 2023, Yahoo! News accessed on 6 July 2023

[156] The Dark Origins of the Davos Great Reset, F. William Engdahl, Global Research, 31 October 2022

[157] The Limits to Growth: A Report for the Club of Rome's Project on the Predicament of Mankind, Donella H. Meadows, Universe Books, 1972, Page 9

[158] The Limits to Growth: A Report for the Club of Rome's Project on the Predicament of Mankind, Donella H. Meadows, Universe Books, 1972, Page 11-12

[159] The Limits to Growth: A Report for the Club of Rome's Project on the Predicament of Mankind, Donella H. Meadows, Universe Books, 1972, Page 186

[160] The Limits to Growth: A Report for the Club of Rome's Project on the Predicament of Mankind, Donella H. Meadows, Universe Books, 1972, Page 187

[161] The Limits to Growth: A Report for the Club of Rome's Project on the Predicament of Mankind, Donella H. Meadows, Universe Books, 1972, Page 23-24

[162] The Limits to Growth: A Report for the Club of Rome's Project on the Predicament of Mankind, Donella H. Meadows, Universe Books, 1972, Pages 190-195

[163] The Limits to Growth: A Report for the Club of Rome's Project on the Predicament of Mankind, Donella H. Meadows, Universe Books, 1972, Pages 72

[164] Limits to Growth: The 30-Year Update, Donella Meadows, Jorgen Randers and Dennis Meadows, Chelsea Green Publishing Company, 2004, Page xvii

[165] The Limits to Growth: A Report for the Club of Rome's Project on the Predicament of Mankind, Donella H. Meadows, Universe Books, 1972, Page 81

[166] Mankind at the Turning Point, The Club of Rome, 1974, Page 69

[167] Mankind at the Turning Point, The Club of Rome, 1974, Page 111

[168] Limits to Growth: The 30-Year Update, Donella Meadows, Jorgen Randers and Dennis Meadows, Chelsea Green Publishing Company, 2004, Page xvii

[169] Report of the United Nations Conference on the Human Environment, Stockholm, 5-16 June 1972, Pages 18-19 accessed on 23 June 2023

[170] Opening statement to the Rio Summit (3 June 1992), MauriceStrong.net accessed on 26 June 2023

[171] The Wizard of Baca Grande, Daniel Wood, West, May 1990, Page 47, captured on 2 August 2014 https://web.archive.org/web/20140802220453/http://www.theageoftransitions.com/images/stories/documents/wizard_baca_grande_1990.pdf and accessed on 23 June 2023

[172] IPCC website, www.ippc.ch accessed on 25 May 2023

[173] Hot Talk, Cold Science: Global Warming's Unfinished Debate, Third Edition, S. Fred Singer with David R Legates and Anthony R. Lupo, Independent Institute, 2021, Chapter 4

[174] Climate Change: The IPCC Scientific Assessment, Edited by J.T.Houghton, G.J.Jenkins and J.J.Ephraums, Intergovernmental Panel on Climate Change, Page 202 downloaded on 29 June 2023

[175] IPCC, 2001: Climate Change 2001: The Scientific Basis. Contribution of Working Group I to the Third Assessment Report of the Intergovernmental Panel on Climate Change [Houghton, J.T., Y. Ding, D.J. Griggs, M. Noguer, P.J. van der Linden, X. Dai, K. Maskell, and C.A. Johnson (eds.)]. Cambridge University Press, Cambridge, United Kingdom and New York, NY, USA, 881pp., Summary for Policymakers, Page 3 downloaded on 29 June 2023

[176] Hot Talk, Cold Science: Global Warming's Unfinished Debate, Third Edition, S. Fred Singer with David R Legates and Anthony R. Lupo, Independent Institute, 2021, Pages 71-73

[177] Fourth Assessment Report (AR4), Intergovernmental Panel on Climate Change (IPCC), Page 468 accessed on 30 May 2023

[178] Corrections to the Mann et. al. (1998) Proxy Data Base and Northern Hemispheric Average Temperature Series, Stephen McIntyre and Ross McKitrick, Energy & Environment, Volume 14, Issue 6, 1 November 2003 accessed on 26 May 2023

[179] Hot Talk, Cold Science: Global Warming's Unfinished Debate, Third Edition, S. Fred Singer with David R Legates and Anthony R. Lupo, Independent Institute, 2021, Page 73-74

[180] Hot Talk, Cold Science: Global Warming's Unfinished Debate, Third Edition, S. Fred Singer with David R Legates and Anthony R. Lupo, Independent Institute, 2021, Chapter 8

[181] Climategate II: More Smoking Guns From The Global Warming Establishment, Larry Bell, Forbes, 29 November 2011 accessed on 25 May 2023

[182] Hot Talk, Cold Science: Global Warming's Unfinished Debate, Third Edition, S. Fred Singer with David R Legates and Anthony R. Lupo, Independent Institute, 2021, Page 78

[183] Hot Talk, Cold Science: Global Warming's Unfinished Debate, Third Edition, S. Fred Singer with David R Legates and Anthony R. Lupo, Independent Institute, 2021, Page 83

[184] Limits to Growth: The 30-Year Update, Donella Meadows, Jorgen Randers and Dennis Meadows, Chelsea Green Publishing Company, 2004, Page xiii-xv

[185] Limits to Growth: The 30-Year Update, Donella Meadows, Jorgen Randers and Dennis Meadows, Chelsea Green Publishing Company, 2004, Page 9

[186] Limits to Growth: The 30-Year Update, Donella Meadows, Jorgen Randers and Dennis Meadows, Chelsea Green Publishing Company, 2004, Page 11

[187] Median income, Wikipedia accessed on 7 July 2023

[188] Limits to Growth: The 30-Year Update, Donella Meadows, Jorgen Randers and Dennis Meadows, Chelsea Green Publishing Company, 2004, Page 261-262

[189] Limits to Growth: The 30-Year Update, Donella Meadows, Jorgen Randers and Dennis Meadows, Chelsea Green Publishing Company, 2004, Page 272-284

[190] Limits and Beyond: 50 years on from The Limits to Growth, what did we learn and what's next, Edited by Ugo Bardi and Carlos Alvarez Pereira, Exapt Press, 2022, Chapter 9, Dr. Ndidi Nnoli=Edozien, Page 122

[191] Limits and Beyond: 50 years on from The Limits to Growth, what did we learn and what's next, Edited by Ugo Bardi and Carlos Alvarez Pereira, Exapt Press, 2022, Chapter 9, Dr. Ndidi Nnoli=Edozien, Pages 122-123

[192] Limits and Beyond: 50 years on from The Limits to Growth, what did we learn and what's next, Edited by Ugo Bardi and Carlos Alvarez Pereira, Exapt Press, 2022, Chapter 9, Dr. Ndidi Nnoli=Edozien, Page 123

[193] Limits and Beyond: 50 years on from The Limits to Growth, what did we learn and what's next, Edited by Ugo Bardi and Carlos Alvarez Pereira, Exapt Press, 2022, Chapter 9, Dr. Ndidi Nnoli=Edozien, Page 124

[194] Limits and Beyond: 50 years on from The Limits to Growth, what did we learn and what's next, Edited by Ugo Bardi and Carlos Alvarez Pereira, Exapt Press, 2022, Chapter 9, Dr. Ndidi Nnoli=Edozien, Page 124

[195] Limits and Beyond: 50 years on from The Limits to Growth, what did we learn and what's next, Edited by Ugo Bardi and Carlos Alvarez Pereira, Exapt Press, 2022, Chapter 9, Dr. Ndidi Nnoli=Edozien, Page 125

[196] Limits and Beyond: 50 years on from The Limits to Growth, what did we learn and what's next, Edited by Ugo Bardi and Carlos Alvarez Pereira, Exapt Press, 2022, Chapter 9, Dr. Ndidi Nnoli=Edozien, Page 130

[197] About us, The Club of Rome website, accessed on 22 June 2023

[198] The Dark Origins of the Davos Great Reset, F. William Engdahl, Global Research, 31 October 2022

[199] Testimony of John R. Christy, U.S. House Committee on Science, Space & Technology, Figure 2, 29 March 2017 downloaded on 25 May 2023

[200] Testimony of John R. Christy, U.S. House Committee on Science, Space & Technology, Section (1), 29 March 2017 downloaded on 25 May 2023

[201] Global Mean Estimates based on Land and Ocean Data, NASA/Goddard Institute for Space Studies, 2019 downloaded on 25 May 2023

[202] Hot Talk, Cold Science: Global Warming's Unfinished Debate, Third Edition, S. Fred Singer with David R Legates and Anthony R. Lupo, Independent Institute, 2021, Page 75

[203] The Sun and Sunspots, National Weather Service, National Oceanic and Atmospheric Administration (NOAA) accessed on 25 June 2024

[204] Historical solar cycles, Space Weather Live accessed on 25 June 2024

[205] Past Temperatures Directly from the Greenland Ice Sheet, D. Dahl-Jensen, K. Mosegaard, N. Gundestrup, G. D. Clow, S. J. Johnsen, A. W. Hansen, and N. Balling, Science, October 1998, Volume 282, Pages 268-271 downloaded on 17 June 2023

[206] All forcing agents CO_2 equivalent concentration, Wikipedia downloaded on 28 April 2023

[207] Scenario Process for AR5, International Panel for Climate Change Data Centre accessed 30 April 2023

[208] "Clean" Energy Exploitations: Helping Citizens Understand the Environmental and Humanity Abuses that Support "Clean" Energy, Ronald Stein / Todd Royal, Archway Publishing, 2021, Page 98

[209] "Clean" Energy Exploitations: Helping Citizens Understand the Environmental and Humanity Abuses that Support "Clean" Energy, Ronald Stein / Todd Royal, Archway Publishing, 2021, Page 98-99

[210] Global Warming, Home Runs, and the Future of America's Pastime, Christopher W. Callahan, et. al., Bulletin of the American Meteorological Society, Volume 104, Issue 5, Page E1010 published 1 May 2023 accessed on 21 May 2023

[211] CO_2: The Greatest Scientific Scandal of Our Time, Zbigniew Jaworowski, EIR Science, 16 March 2007 accessed on 28 June 2023

[212] CO_2: The Greatest Scientific Scandal of Our Time, Zbigniew Jaworowski, EIR Science, 16 March 2007, Page 44 accessed on 28 June 2023

[213] CO_2: The Greatest Scientific Scandal of Our Time, Zbigniew Jaworowski, EIR Science, 16 March 2007, Page 45 accessed on 28 June 2023

[214] CO_2: The Greatest Scientific Scandal of Our Time, Zbigniew Jaworowski, EIR Science, 16 March 2007 accessed on 29 June 2023, Page 46-47

[215] CO_2: The Greatest Scientific Scandal of Our Time, Zbigniew Jaworowski, EIR Science, 16 March 2007 accessed on 29 June 2023, Page 47

[216] Analysis: Dark summer nights: India faces high risks of power cuts after years of coal, hydro power neglect, Sudarshan Varadhan, Sarita Chaganti Singh and Matthew Chye, Reuters, 9 March 2023 accessed on 12 June 2023

[217] U.S. Product Supplied of Finished Motor Gasoline, U.S. Energy Information Administration downloaded on 19 May 2023

[218] U.S. All Grades All Formulations Retail Gasoline Prices, U.S. Energy Information Administration downloaded on 19 May 2023

[219] A wealth tax could help poorer countries tackle climate crisis, economists say, Fiona Harvey, The Guardian, 19 January 2023 accessed on 19 June 2023

[220] Global carbon dioxide removal rates from forest landscape restoration activities, Blanca Bernal, Lara T. Murray and Timothy R. H. Pearson, Carbon Balance and Management, Volume 13, Article Number: 22, 2018 accessed on 12 April 2023

[221] File:"You'll Die of Old Age I'll Die of Climate Change" - Climate Strike Toronto, 27 Sep 2019, Wiki Commons downloaded on 22 May 2023

[222] Recycling Facts, Middle Georgia State University accessed on 2 April 2023

[223] Aluminum recycling, Wikipedia accessed on 2 April 2023

[224] Most recycled material in the world, The World Counts, accessed on 2 April 2023

[225] Paper recycling facts and statistics, World Economic Forum accessed on 2 April 2023

[226] Plastic pollution is growing relentlessly as waste management and recycling fall short, Organisation for Economic Co-operation and Development (OECD) accessed on 2 April 2023

[227] Plastic Recycling Doesn't Work and Will Never Work, Judith Enck, Jan Dell, The Atlantic, 30 May 2022 accessed on 4 December 2023

[228] HERE study: The French and the proposal of a quota of 4 flights per person in a lifetime, Consumer Science & Analytics, 28 July 2023 accessed on 3 November 2023

[229] Direct air capture's hidden energy cost, Michelle Ma, Protocol, 21 October 2022 accessed on 14 January 2024

[230] Energy Fundamentals of Carbon Capture, Ben James, Share Your Green Design, 2 August 2023 accessed on 14 January 2024

[231] Freedonia Report Finds New Jersey Single-Use Bag Ban Boosts Alternative Bag Production, Increases Plastic Consumption, and Drives Retailer Profits, by Kristen Pieffer, Freedonia Custom Research, 9 January 2024 accessed on 25 January 2024

[232] Freedonia Report Finds New Jersey Single-Use Bag Ban Boosts Alternative Bag Production, Increases Plastic Consumption, and Drives Retailer Profits, by Kristen Pieffer, Freedonia Custom Research, 9 January 2024 accessed on 25 January 2024

[233] What is U.S. electricity generation by energy source?, U.S. Energy Information Administration accessed on 16 April 2023

[234] Figure 8-1, What is U.S. electricity generation by energy source?, US Energy Information Administration, and various locations on the Statista website accessed on 31 March 2023

[235] India's pledge to stop new coal power plants to hit key states, Gavin Maguire, 9 May 2023, Reuters accessed on 13 June 2023

[236] China Pledges to Stop Building Coal-Burning Power Plants Abroad, Somini Sengupta and Rick Gladstone, New York Times, updated 28 October 2021 accessed on 3 April 2023

[237] China permits two new coal power plants per week in 2022, Centre for Research on Energy and Clean Air (CREA) accessed on 3 April 2023

[238] 90% of All the Scientists That Ever Lived Are Alive Today, Eric Gastfriend, Future of Life Institute, 5 November 2015 accessed on 18 June 2023

[239] Dow's Corpus Christi project highlights challenge of nuclear energy's revival, James Osborne, Houston Chronicle, 1 December 2023 accessed on 3 January 2024

[240] Chemical Processing Notebook: Will Mini Nuclear Plants Power the Chemical Industry's Future?, Jonathon Katz, Chemical Processing, 13 November 2023 accessed on 17 November 2023

[241] Chemical Processing Notebook: Will Mini Nuclear Plants Power the Chemical Industry's Future?, Jonathon Katz, Chemical Processing, 13 November 2023 accessed on 17 November 2023

[242] X-energy website https://x-energy.com/fuel/triso-x

[243] Controlling the Future - Controlling Nonindustrial Processes: Preventing Climate and Other Disasters. Bela Liptak, ISA Press, 2023, Page xvi

[244] How does the land use of different electricity sources compare?, Hannah Ritchie, 16 June 2022, Our World in Data accessed on 10 April 2023

[245] How Much CO_2 Does a Tree Absorb?, One Tree Planted, Ross Bernet, 5 October 2021 accessed on 10 April 2023

[246] How Many Birds Are Killed by Wind Turbines?, Joel Merriman, 26 January 2021, American Bird Conservatory accessed on 4 April 2023

[247] "Clean" Energy Exploitations: Helping Citizens Understand the Environmental and Humanity Abuses that Support "Clean" Energy, Ronald Stein / Todd Royal, Archway Publishing, 2021, Page 185

[248] NOAA: N.J. wind farm may 'adversely affect,' not kill whales, WHYY, 4 April 2023 accessed on 3 May 2023

[249] "Clean" Energy Exploitations: Helping Citizens Understand the Environmental and Humanity Abuses that Support "Clean" Energy, Ronald Stein / Todd Royal, Archway Publishing, 2021, Chapter Five

[250] Lithium-ion battery fires are happening more often. Here's how to prevent them, Samantha Murphy Kelly, CNN Business, 9 March 2023 accessed on 20 June 2023

[251] Controlling the Future - Controlling Nonindustrial Processes: Preventing Climate and Other Disasters. Bela Liptak, ISA Press, 2023

[252] Closed-loop CO_2-based energy-storage system slated for Wisconsin, Chemical Engineering, November 2023, Page 6

[253] European travel strikes in April and May: When are they and what can you expect?, updated on 4 November 2022, euronews.travel accessed on 11 April 2023

[254] Why EV charging is still such a pain, Peter Valdes-Dapena, CNN, 4 November 2023 accessed on 20 November 2023

[255] City/highway mileages for Toyota Corolla, Toyota Corolla Hybrid, and Toyota bZ4X, fueleconomy.gov accessed on 4 April 2023

[256] "Clean" Energy Exploitations: Helping Citizens Understand the Environmental and Humanity Abuses that Support "Clean" Energy, Ronald Stein / Todd Royal, Archway Publishing, 2021, Chapter Five

[257] "Clean" Energy Exploitations: Helping Citizens Understand the Environmental and Humanity Abuses that Support "Clean" Energy, Ronald Stein / Todd Royal, Archway Publishing, 2021, Page 124

[258] 2022 Estimates from Statista accessed on 1 May 2023

[259] The new 'gold rush' for green lithium, Catherine Early, Future Planet, 24 November 2020 accessed on 11 April 2023

[260] How much CO_2 is emitted by manufacturing batteries?, Iris Crawford, MIT MechE, 1 March 2022 accessed on 11 April 2023

[261] How Much Does an EV Battery Cost to Replace?, Recurrent accessed on 25 April 2023

[262] Volkswagen Green Finance Report, Volkswagen, July 2022 downloaded on 11 April 2023

[263] State Air Board Repeals Mandate for Electric Cars, Marla Cone, Los Angeles Times, 30 March 1996 accessed on 6 April 2023

[264] DOE Finalizes Energy Efficiency Standards for Residential Furnaces to Save Americans $1.5 Billion In Annual Utility Bills, 29 September 2023, energy.gov accessed on 6 October 2023

[265] Experts warn Biden admin's water heater crackdown will hike prices, reduce consumer choice, Thomas Catenacci, Fox News, 26 July 2023 accessed on 28 July 2023

[266] Experts warn Biden admin's water heater crackdown will hike prices, reduce consumer choice, Thomas Catenacci, Fox News, 26 July 2023 accessed on 28 July 2023

[267] Food's Carbon Footprint, Green Eatz downloaded on 12 April 2023

[268] Your carbon footprint may have more to do with your wealth than your location, Sarah DeWeerdt, Anthropocene, 11 February 2020 accessed on 7 April 2023

[269] Distribution of carbon dioxide emissions worldwide in 2021, by select country, Statista, accessed on 20 June 2023

[270] Average Annual Miles per Driver by Age Group, Federal Highway Administration, US Department of Transportation modified on 31 May 2022 and accessed on 5 April 2023

[271] Number of motor vehicles registered in the United States from 1990 to 2021, Statista accessed on 5 April 2023

[272] Variable Speed Drives: Principles and Applications for Energy Cost Savings, David W. Spitzer, Momentum Press, 2013

[273] Will Carbon Dioxide To Propane Conversion Close Carbon Cycle?, Traci Purdum, Chemical Processing, 18 August 2023 accessed on 20 November 2023

[274] Catalyst Breakthroughs: Transforming Plastics and Photocatalysis, Sean Ottewell, 24 October 2023, Chemical Processing, accessed on 20 November 2023

[275] A new electrolyzer system makes propane from CO_2, Gerald Ondrey (Editor), Chemical Engineering, October 2023, Page 6

[276] Scaled up facility under construction for single-step, CO_2-to-fuels process, Gerald Ondrey (Editor), Chemical Engineering, October 2023, Page 6

[277] Carbon Capture, Gerald Ondrey (Editor), Chemical Engineering, October 2023, Page 6

[278] Chemical Looping, Gerald Ondrey (Editor), Chemical Engineering, October 2023, Page 6

[279] Seawater electrolysis stabilizes and immobilizes atmospheric carbon dioxide, Gerald Ondrey (Editor), Chemical Engineering, October 2023, Page 7

[280] Using waste plastic to simultaneously make graphene and hydrogen, Gerald Ondrey (Editor), Chemical Engineering, November 2023, Page 5

[281] Commercial Progress in Turqoise Hydrogen, Scott Jenkins, Carl Fromm, Chemical Engineering, November 2023, Pages 12-16

[282] History of Air Pollution, United States Department of Environmental Protection (EPA) accessed on 5 March 2023

[283] Remotely Set Pressure Regulators: A User's Perspective, David W Spitzer, Intech Magazine, November 1992, Pages 40-43

[284] Oil and petroleum products explained, U. S. Energy Information Administration accessed on 23 May 2023

[285] Primary energy consumption by source (2021), Our World in Data accessed on 21 April 2023

[286] Per capita CO_2 emissions, Our World in Data downloaded on 20 April 2023

[287] Countries in the world by population (2023), Worldometer accessed on 28 April 2023

[288] Annual CO_2 emissions by world region, Our World in Data downloaded on 20 April 2023

[289] "Clean" Energy Exploitations: Helping Citizens Understand the Environmental and Humanity Abuses that Support "Clean" Energy, Ronald Stein / Todd Royal, Archway Publishing, 2021, Chapter Six

[290] "Clean" Energy Exploitations: Helping Citizens Understand the Environmental and Humanity Abuses that Support "Clean" Energy, Ronald Stein / Todd Royal, Archway Publishing, 2021, Page 218

[291] Electricity production by source, Our World in Data downloaded on 20 April 2023

[292] Death rate from disasters, Our World in Data downloaded on 21 July 2023

[293] Jerry Brown's Secret War on Clean Energy, Environmental Progress, 11 January 2018 accessed on 23 May 2023

[294] The Limits to Growth: A Report for the Club of Rome's Project on the Predicament of Mankind, Donella H. Meadows, Universe Books, 1972, Page 23-24

[295] *An Essay on the Principle of Population*, Thomas Malthus, 1798, Page vii

[296] The Limits to Growth: A Report for the Club of Rome's Project on the Predicament of Mankind, Donella H. Meadows, Universe Books, 1972, Pages 190-195

[297] Limits to Growth: The 30-Year Update, Donella Meadows, Jorgen Randers and Dennis Meadows, Chelsea Green Publishing Company, 2004, Page 11

[298] Median income, Wikipedia accessed on 7 July 2023

[299] Mankind at the Turning Point, The Club of Rome, 1974, Page 69

[300] Mankind at the Turning Point, The Club of Rome, 1974, Page 111

[301] Socialism, Meriam-Webster accessed on 30 June 2023

[302] Limits and Beyond: 50 years on from The Limits to Growth, what did we learn and what's next, Edited by Ugo Bardi and Carlos Alvarez Pereira, Exapt Press, 2022, Chapter 9, Dr. Ndidi Nnoli-Edozien, Page 124

[303] Limits and Beyond: 50 years on from The Limits to Growth, what did we learn and what's next, Edited by Ugo Bardi and Carlos Alvarez Pereira, Exapt Press, 2022, Chapter 9, Dr. Ndidi Nnoli-Edozien, Page 124

[304] Limits and Beyond: 50 years on from The Limits to Growth, what did we learn and what's next, Edited by Ugo Bardi and Carlos Alvarez Pereira, Exapt Press, 2022, Chapter 9, Dr. Ndidi Nnoli-Edozien, Page 130

[305] Gas stove manufacturers push back on talk of a ban, Lisa Fickenscher, New York Post, 10 January 2023 accessed on 5 July 2023

[306] Methane and NOx Emissions from Natural Gas Stoves, Cooktops, and Ovens in Residential Homes, Eric D. Lebel, Colin J. Finnegan, Zutao Ouyang, and Robert B. Jackson, Environmental Science & Technology, 56 (4), 2022, Figure 5, Page 2536

[307] Reparations for All or None, Katherine Brush, 2022, Page 326

[308] Singapore, United Nations Fund for Population Activities UNFPA, National Library of Medicine, National Institute of Health accessed on 3 November 2023

[309] Demographic history of Palestine (region), Wikipedia accessed on 3 November 2023

[310] Resource Mismanagement a Threat to Security in Africa, Paul Nantulya, African Center for Strategic Studies, 7 September 2016 accessed on 3 July 2023

[311] Reference to 'teaching a man to fish' isn't from Bible, The Repository, 12 September 2010 accessed on 3 July 2023

[312] History and slavery, Wikipedia accessed on 19 June 2023

[313] *'The world is going to end in 12 years if we don't address climate change,' Ocasio-Cortez says*, USA Today, 22 January 2019

[314] It's Been 5 Years Since Greta Thunberg Warned Climate Change Would 'Wipe Out All of Humanity Unless We Stop Using Fossil Fuels Over the Next 5 Years', Andrew Stiles, The Washington Free Beacon, 21 June 2023 accessed on 27 June 2023

[315] Limits to Growth: The 30-Year Update, Donella Meadows, Chelsea Green Publishing, 2004, Chapter 5

[316] Sample Good Books Related To Anthropogenic Global Warming, wiseenergy.com accessed on 1 May 2023

[317] Climate Change Reconsidered II: Fossil Fuels (2019), Nongovernmental International Panel on Climate Change (NIPCC), Chapter 1.5.3 accessed on 27 May 2023

[318] Climate Change Reconsidered II: Fossil Fuels (2019), Nongovernmental International Panel on Climate Change (NIPCC), Chapter 1.4.3 accessed on 27 May 2023

[319] Executive Order on Protecting Public Health and the Environment and Restoring Science to Tackle the Climate Crisis, Section 1, whitehouse.gov, 20 January 2021 accessed on 1 May 2023

[320] What Is Environmental Justice?, Office of Legacy Management, U. S, Department of Energy accessed on 1 May 2023

[321] Executive Order on Protecting Public Health and the Environment and Restoring Science to Tackle the Climate Crisis, Section 6, whitehouse.gov, 20 January 2021 accessed on 1 May 2023

[322] State Air Board Repeals Mandate for Electric Cars, Marla Cone, Los Angeles Times, 30 March 1996 accessed on 6 April 2023

[323] Study compares electric vehicle charge costs vs. gas — and results were surprising, Jamie L. LaReau, Detroit Free Press, 21 October 2021 accessed on 9 April 2023

[324] Advanced Clean Cars II Regulations: All New Passenger Vehicles Sold in California to be Zero Emissions by 2035, California Air Resources Board, ca.gov accessed on 29 April 2023

[325] US approves California plan requiring half of heavy duty trucks be EV by 2035, David Shepardson, 31 March 2023, Reuters accessed on 29 April 2023

[326] Governor Newsom's Zero-Emission by 2035 Executive Order (N-79-20), 19 January 2021 accessed on 29 April 2023

[327] Electric Generation Capacity and Energy, California Energy Commission accessed on 30 April 2023

[328] California planned to close down its last nuclear plant by 2025 and replace it with clean energy. What went wrong?, Kavya Balaraman, Utility Dive, 9 September 2022 accessed on 29 April 2023

[329] Why a California city is trying to build the state's last fossil-fueled power plant, 5 March 2022, CNBC accessed on 29 April 2023

[330] Summary of California Vehicle and Transportation Energy (October 2015), California Energy Commission accessed on 30 April 2023

[331] Flex Alert extended to Saturday; EV owners asked to not charge vehicles during peak hours, 2 September 2022, CBS News accessed on 30 April 2023

[332] Electric big rigs are going farther and charging faster, Jeff St. John, 25 September 2023, Canary Media accessed on 3 November 2023

[333] Semi: The Future of Trucking, https://www.tesla.com/semi accessed on 3 November 2023

[334] 10 tips to managing semi-truck's fuel efficiency, Hale Trailer Brake & Wheel, 3 February 2023 accessed on 3 November 2023

[335] Opening remarks by NATO Secretary General Jens Stoltenberg on NATO 2030 and the importance of strengthening the transatlantic bond in the next decade and beyond, North Atlantic Treaty Organization (NATO), 4 February 2021 accessed on 5 June 2023

[336] The British Army's Next 'Game-Changer' Weapon: Electric Tanks?, Michael Peck, The National Interest, accessed on 2 November 2019

[337] Local Laws of the City of New York for the Year 2015, Number 38, Subchapter 2, Section 24-105

[338] Climate Change Reconsidered II: Fossil Fuels (2019), Nongovernmental International Panel on Climate Change (NIPCC), Chapter 7.1.1 accessed on 27 May 2023

[339] Congressional Bills 116th Congress, U.S. Government Publishing Office, S. Res. 59 Introduced in Senate (IS), 116th CONGRESS, 1st Session, S. RES. 59, 7 February 2019

[340] Congressional Bills 116th Congress, U.S. Government Publishing Office, S. Res. 59 Introduced in Senate (IS), 116th CONGRESS, 1st Session, S. RES. 59, 7 February 2019

[341] This prairie chicken is Biden's latest weapon in his war on fossil fuels, Kris Kobach, Fox News, 2 May 2023 accessed on 2 May 2023

[342] This prairie chicken is Biden's latest weapon in his war on fossil fuels, Kris Kobach, Fox News, 2 May 2023 accessed on 2 May 2023

[343] This prairie chicken is Biden's latest weapon in his war on fossil fuels, Kris Kobach, Fox News, 2 May 2023 accessed on 2 May 2023

[344] False Alarm: How Climate Change Costs Us Trillions, Hurts the Poor, and Fails to Fix the Planet, Bjorn Lomborg, Basic Books, Page 214

[345] "Clean" Energy Exploitations: Helping Citizens Understand the Environmental and Humanity Abuses that Support "Clean" Energy, Ronald Stein / Todd Royal, Archway Publishing, 2021, Page 2

[346] Jane Fonda Climate PAC, www.janepac.com accessed on 29 May 2023

[347] Jane Fonda Climate PAC, www.janepac.com accessed on 29 May 2023

[348] Jane Fonda says climate crisis can be blamed on racism: 'Where would they put the sh—? Not Bel Air.', MarketWatch, 28 January 2023 accessed on 29 May 2023

[349] Maurice Strong, Wikipedia accessed on 24 June 2023

[350] "WHO IS MAURICE STRONG? The adventures of Maurice Strong & Co. illustrate the fact that nowadays you don't have to be a household name to wield global power", National Review, September 1, 1997

[351] The Wizard of Baca Grande, Daniel Wood, West, May 1990, Page 29, captured on 2 August 2014 at https://web.archive.org/web/20140802220453/http://www.theageoftransitions.com/images/stories/documents/wizard_baca_grande_1990.pdf and accessed on 24 June 2023

[352] Global cooling, Wikipedia accessed on 7 July 2023

[353] Limits and Beyond: 50 years on from The Limits to Growth, Edited by Ugo Bardi and Carlos Alvarez Pereira, How the Club of Rome influenced the world's agenda, Wouter van Dieran, Exapt Press, 2022, Page 102

[354] Less burning, no tears, R L Stanfield, National Journal, 13 August 1988, Pages 2095-2098 (Attribution)

[355] The First Global Revolution, Alexander King and Bertrand Schneider, The Club of Rome, 1991, Page 115 accessed on 23 June 2023

[356] Natural Resources Defense Council, Consolidated Statement of Activities for the year ended June 30, 2022, Natural Resources Defense Council website accessed on 8 July 2023

[357] Support NRDC'S Work for the Environment, National Resources Defense Council website accessed on 7 July 2023

[358] Health Scares That Weren't So Scary, Jane E Brody, New York Times, 18 August 1998 accessed on 8 July 2023

[359] How a PR Firm Executed the Alar Scare, Wall Street Journal, 3 October 1989, Page A22:4 accessed on 7 July 2023

[360] State Air Board Repeals Mandate for Electric Cars, Marla Cone, Los Angeles Times, 30 March 1996 accessed on 6 April 2023

[361] The Sustainable Development Agenda, United Nations accessed on 12 July 2023

[362] Sustainable Development Goals, Wikipedia accessed on 12 July 2023

[363] An Essay on the Principle of Population, Thomas Malthus, 1798, Page vii

[364] Sustainable Development Goals, Wikipedia accessed on 12 July 2023

[365] Apple Growers Bruised and Bitter After Alar Scare, Timothy Egan, The New York Times, 9 July 1991 accessed on 5 January 2023

[366] How a PR Firm Executed the Alar Scare, Wall Street Journal, 3 October 1989, Page A22:4 accessed on 7 July 2023

[367] An Essay on the Principle of Population, Thomas Malthus, 1798, Page vii